AF689464

Revolutionizing Higher Education in Agriculture

Revolutionizing Higher Education in Agriculture

Framework, Principles, and Agenda for Action

Edited by H. O. Kunkel and C. L. Skaggs

Iowa State University Press / Ames

H. O. Kunkel, Professor Emeritus, Animal Science, and Dean Emeritus, College of Agriculture and Life Sciences, Texas A&M University.

C. L. Skaggs, Associate Professor, Animal Science, Texas A&M University.

This work was supported in part by Challenge Grants from Higher Education Programs, Science and Education Resources Development, Cooperative State Research, Extension and Education Service, United States Department of Agriculture.

© 2001 Iowa State University Press
All rights reserved.

Orders: 1-800-862-6657
Office: 1-515-292-0140
Fax: 515-292-3348
Web site: www.isupress.com

Authorization to photocopy items for internal or personal use, or the internal or personal use of specific clients, is granted by Iowa State University Press, provided that the base fee of $.10 per copy is paid directly to the Copyright Clearance Center, 222 Rosewood Drive, Danvers, MA 01923. For those organizations that have been granted a photocopy license by CCC, a separate system of payments has been arranged. The fee code for users of the Transactional Reporting Service is 0-8138-1774-9/2001$.10.

♾ Printed on acid-free paper in the United States of America

First edition, 2001

Library of Congress Cataloging-in-Publication Data

Revolutionizing higher education in agriculture : framework, principles and agenda for action / edited by H. O. Kunkel and C. L. Skaggs.
p. cm.
Includes bibliographical references (p.).
ISBN 0-8138-1774-9 (alk. paper)
1. Agriculture—Study and teaching (Higher)—United States. 1. Kunkel, H. O. II. Skaggs, C.L.

S533.W68 2001
630′.71′173—DC21 00–046136

The last digit is the print number: 9 8 7 6 5 4 3 2 1

Contents

Contributors

Work Group on Systemic Change in Undergraduate Education in Agriculture, Phase One

John Brand, University of Connecticut

George Brown, Alabama A&M University

Willie Cheatham, Alabama A&M University; Project Codirector

George Conneman, Cornell University

William Drake, Cornell University

Cameron Faustman, University of Connecticut

McArthur Floyd, Alabama A&M University

Hugo John, University of Connecticut

H. O. Kunkel, Texas A&M University; Project Director

Ian L. Maw, Cook College, Rutgers University

William H. Neill, Texas A&M University

A. Gene Nelson, Texas A&M University

John Riesen, University of Connecticut

James Shuford, Alabama A&M University

Suman Singha, University of Connecticut

C. L. Skaggs, Texas A&M University; Project Director after February1998

Gopal Sunki, Alabama A&M University

Dean Sutphin, Cornell University

Paul B. Thompson, Texas A&M University

Christine Townsend, Texas A&M University

Donald Vietor, Texas A&M University

Faculty Consultants

Phil Arneson, Cornell University

J. Bafacadu, Alabama A&M University

Ron Butler, Cornell University

H. Jones, Alabama A&M University

Wayne Knoblauch, Cornell University

O. G. Sanders, Alabama A&M University

Harry Stinson, Cornell University

Dwayne Suter, Texas A&M University

Bruce Wilkins, Cornell University

Student Consultants

Sharla Adams, Texas A&M University

Trace Blair, Texas A&M University

Gregory Chovanec, Texas A&M University

Brian Markwardt, Texas A&M University

Andrew Murphey, Texas A&M University

Tonnyre Thomas, Texas A&M University

Bonnie Woolam, Texas A&M University

Core Work Group on Implementation of Systemic Change in Undergraduate Education in Agriculture, Phase Two (Current Positions)

McArthur Floyd, Associate Dean for Research, School of Agriculture and Environmental Sciences, Alabama A&M University

H. O. Kunkel, Professor Emeritus, Animal Science, and Dean Emeritus, College of Agriculture and Life Sciences, Texas A&M University; Project Director

Ian L. Maw, Dean of Academic and Student Affairs, Cook College, Rutgers University

William H. Neill, Professor, Wildlife and Fisheries Science, Texas A&M University

A. Gene Nelson, Professor and Head, Agricultural Economics, Texas A&M University

Suman Singha, Associate Dean, College of Agriculture and Natural Resources, University of Connecticut

C. L. Skaggs, Associate Professor, Animal Science, Texas A&M University, Project Director after February 1998

H. Dean Sutphin, Associate Dean and Director of Academic Programs, College of Agriculture and Life Sciences, Cornell University.

Paul B. Thompson, Professor, Department of Philosophy, Purdue University

Donald Vietor, Professor, Soil and Crop Sciences, Texas A&M University

Preface

As the study of concepts of change in higher education in agriculture and related areas evolved, several facts became evident. The entire food, agricultural, and natural resources scene has changed in the last decade. Congress has written a farm bill that for the first time in history is not fashioned in the agrarian tradition. The market system for much of agriculture, particularly animal agriculture and horticultural crops, has been restructured, involving centralized processing plants, corporate supply of transport, and distribution through corporate-controlled outlets. Much of the production is industrialized. The job market for graduates requires increasing flexibility but, in fact, seems to be growing. The educational system serves a much broader constituent base. Far more individuals than those of the traditional constituency have interests in food, agriculture, and the natural resources.

The system of higher education in agriculture and related areas is not a trivial system. It has vitality, importance to society, and intrinsic values as a body of knowledge. But its external and internal environments are undergoing massive changes. The system is asked to respond.

Increasing numbers of institutions of higher education in agriculture and related areas have been individually reexamining their programs and engaging in efforts to revitalize their programs. The growing awareness of the importance of systemic (meaning comprehensive, fundamental) change to the continued relevance and vitality of higher education in agriculture has created a need for both understanding of the theoretical bases of agricultural higher education and for suggestions and guidance. To begin to meet this need, a USDA Higher Education Programs Challenge Grant to Texas A&M University and Alabama A&M University, entitled Theoretical Bases of Systemic Change in Higher Education in Agriculture, was activated 1 September 1993. A second Challenge Grant, entitled Strategies for Implementation of Systemic Changes in Higher Education in Agriculture, was awarded to Texas A&M University on 1 September 1994. The stated objectives were (1) to develop theoretical bases and propose principles for systemic change in undergraduate education in agriculture, (2) to seek validation of the proposed guidelines, and (3) to provide guidance for implementation of change in colleges of agriculture that wish to make such changes. The first edition of this book was directed at the first objective, the development of concepts for systemic change, and provides a basis for movement toward further, ongoing objectives of the project. This second edition extends across all three objectives.

In the initial phase, faculty members of five universities, Alabama A&M University, University of Connecticut, Cornell University, Rutgers University, and Texas A&M University, were brought together to engage in the process of discovery and development of applicable theoretical principles. Such principles and theories are expected to lead to goals that individual institutions might use to implement change. If accepted by the agricultural higher education community, such principles and theories might be used as bases for the general reconstruction of higher education in agriculture in the twenty-first century. It follows that, if new principles are accepted by the agricultural higher education community, then strategies for implementation need to be created and adopted.

One purpose of this effort is to promote articulation of a variety of views to strengthen the cogency and coherence of the several arguments. All too often, positions on higher education rely on an assortment of "illustrative examples" or deeply held convictions that are scantily supported and weakly argued. That is not enough. It is more important than ever that an interdisciplinary pluralism of perspectives be built and that we learn from it.

The principles and theories for systemic change in higher education in agriculture will need continued refinement, debate, and discussion. So will the strategies for implementation.

What is ultimately important, however, is that there be a sense of ownership by the faculties in the individual colleges, schools, and departments of agriculture. It is to initiate that process that this effort was undertaken.

The twenty-member, initial Work Group on Systemic Change in Undergraduate Education in Agriculture was designated by the respective offices of the deans of agriculture at the cooperating institutions. Focus groups of students were selected by members of the Work Group. Collectively, the membership consisted of faculty in a multiplicity of disciplines, including agricultural and resource economics, agricultural education, agronomy, soils, animal sciences, food sciences, horticulture, forestry and other natural resources, wildlife and fisheries sciences, ethics and policy, and human nutrition. The same range of disciplines characterized the faculty members brought together for workshops that focused on implementation. In the preparations of their contributions, the participants drew on their own resources as well as those of others whose contributed knowledge was integrated in the thought processes. All members of the Work Group engaged in discussions on their own campuses and at the workshops at Washington, D.C., and College Station, Texas. A smaller number of the group participated in the implementation workshops in College Station, Texas, and Ithaca, New York, in 1997 and 1998, respectively. Significantly, a restated generic purpose of colleges of agriculture was accepted by the group.

This book consists of two parts: (1) the framework and (2) the focus on implementation.

The chapters in part 1 consist of theoretical and conceptual analyses of the issues and underpinnings of higher education in agriculture. In these chapters, the contributing authors express judgments that may seem similar, but each is framed differently in the context of the chapter. These chapters are essays that should contribute to the understanding of the purpose and essence of higher education in agriculture and its kindred subject areas and the process of systemic change.

Part 2 consists of the examination of strategies for implementation of the changes suggested by theory and principle. These chapters are derived from the two national implementation workshops, in which different approaches to implementation were expressed by plenary speakers and were discussed by the participants. They build upon and add to the theories and principles of the first phase.

This book presents concepts. It does not provide a blueprint for carrying out individual institutional systemic change, although it has a closing agenda for action. The recommendations are general by design. They are intended to be primarily part of the process of the rethinking needed to bring higher education in agriculture into the twenty-first century, as a viable and necessary component of higher education in the United States.

The future is clearly in the minds of the contributors and participants of the study. An early review of the manuscript noted that the concepts developed in this study are also in a constant state of accelerating change. Technology, information generation and use, the understanding of learning process, and social values are all changing. We agree. Perhaps, it would have been better, if possible, to explore the expected future change of the context and concepts that were established in this study. It is clearly evident that the complexities of the biological, economic, and social bases of agriculture and natural resources will be far greater in the twenty-first century than any time in the past. The future is uncertain. But the fact is noted that the concepts that guide higher education are also unstable.

Having largely met the challenge of educating young professionals in a fairly well defined food and agriculture system over the past decades, colleges of agriculture must now face the even greater challenge of education of young men and women to be flexible enough to find profitable work in the great diversity of careers related to food, agriculture, and the human environment that will be the norm of the future in an environment that we know will be increasingly complex. To the extent that this report leads to efforts to face and meet this challenge through discussion, debate, new initiatives, and extension of initiatives already in progress, the efforts will be considered to be worthwhile.

This report is a team effort. The participants in the studies were bold thinkers. They were imaginative and innovative. They were dedicated to the undergraduate students, to research and outreach, and to building the strength and vitality of the system of higher education in agriculture, food, and the natural resources. As editors, we do not change what they have said.

Our thanks go to all of the individuals who contributed to this report by providing important information and unique views of higher education. We are especially grateful to Dr. K. Jane Coulter of Science and Education Resources Development, Cooperative State Research, Extension and Education Service, USDA, for her interest and interaction with us. We are grateful to Dean Edward Hiler, Texas A&M University, Dean David Call, Cornell University, Interim Dean John Brand, University of Connecticut, and Dean James Shuford, Alabama A&M University, for their early support of the concept of the project.

Finally, we thank Kathy Stasny and Lela Batcheler, Department of Animal Science, Texas A&M University, for office help, arrangements, and typing and producing multiple copies of the many versions of the manuscript.

H. O. Kunkel
C. L. Skaggs

How to Use This Book

It is easy to be cynical about systemic change. One idea is that reports such as this serve as mandates for action. Administrators and legislators may act on the summary bullets for which the body of the report is assumed to provide in-depth analysis and support. But this form of action will not likely go beyond incremental improvement or adjustment. If systemic change in agricultural colleges is brought about from within, it will not be based upon argument so thin that it can be summarized in bullets. It will not be brought about by people too busy to read a book or to spend time talking and arguing with their colleagues, and a corollary is that it will not happen at colleges where the institutional values do not support the commitment of faculty time for the task of collectively formulating an institutional vision. At such colleges, systemic change will come, but it will be imposed from without.

Systemic change will come from within when key people, administrators and faculty, invest time in arriving at a common language and an agreement that there is a problem. They need not agree on the solutions, but a common language is necessary in order to assure that disagreements are substantive, rather than terminological. Given the specialization that is the norm in our universities, however, this common language will not come automatically, or even easily, but will require a significant investment of time and energy. Students should be included as well. Educational institutions will need to encourage the contributions of students to decision making concerning education.

Those in the key group will have to frame a vision collaboratively. A way toward one's own vision is to first consider the vision of others. The key group must do this in seminar fashion (which is, after all, the way it was done among faculties for centuries). This book can facilitate change more effectively if it is used as a prod or a touchstone for a careful and painstaking group reflection and debate than if it is used as a mandate for administrative decision. Used in this fashion, it will matter less that what has been said here is applicable to a particular setting than that these ideas provide a basis for the conversations that will build a common language and a common sense of the problem definition. There is also a larger aspect that transcends the individual college of agriculture. New curricular and teaching methodologies and resources might be led by the professional system, including the land grant system, and the professional and learned societies. Regional symposia and conferences draw faculty from all disciplines, mainly for the purpose of improving the skills and resources of individual faculty members. Such regional conferences can become increasingly concerned with institutional improvement. Here, too, a common language is needed. This book, again, can facilitate change if used in reflection and debate.

Revolutionizing Higher Education in Agriculture

Part One

The Framework and Principles for Changes

Since the publication of the report of the 1991 USDA-National Research Council (NRC) national conference on professional education of the undergraduate, "Agriculture and the Undergraduate" (NRC, 1992), efforts have significantly expanded to build new models for higher education in the agricultural sciences and their closely related areas. Occurring is healthy rethinking of education in agriculture and natural resources as well as of the broader missions of the land grant colleges of agriculture, food, and natural resources.

Change is imperative. The complexities of agriculture, food systems, and natural resources demand that the content and context of undergraduate and professional education reflect the diversity of needs in graduates in the future. Add to this the revolutions in some of the underlying sciences and technologies: economics, ecology, molecular biology, physiology, computer technology, biotechnology, and more. The traditional approaches can no longer suffice. Integration and hierarchical considerations are essential.

At the same time that this study developed, a number of efforts were underway to develop the model(s) for the new generations of higher education in agriculture and related areas. The National Research Council undertook a study of the future of the land grant colleges of agriculture (NRC, 1996). The W. K. Kellogg Foundation sponsored a nationwide initiative for development of the food systems professional education by the year 2020. The Board on Agriculture of the National Association of State Universities and Land Grant Colleges (NASULGC) developed a system of listening and synthesis efforts to complement the Kellogg initiative and built a consensus report (Carpenter and Fisher, 1996). In addition, James H. Meyer, who lists this study (Kunkel et al., 1996) as also one that reflected national concerns, has written extensively on the past and future of the land grant college of agriculture (Meyer 1992, 1993, 1997, 1998).

The common theme is that the need for development and implementation of the concepts and framework for change has become clear. As the models for higher education in agriculture and related areas are developed, the institution in which change is implemented also requires conceptual description. This study of the theory and implementation of systemic change in higher education in agriculture focuses on that need.

This book differs from other studies: The emphasis is on undergraduate education. It focuses on the place and on the people who will make the change, namely, the faculty and administration of the college or department, and develops a framework for making the changes for which the institution sees the need. Throughout this book the term *agriculture* is used to denote all of the traditional aspects of the food and fiber system, including but not limited to agriculture, food, agribusiness, forestry, wildlife, recreation, fisheries, range, water, and environment; the continuum from food and fiber production to consumption; and the scientific, educational, and governmental infrastructure. The term *college of agriculture* is to denote generically colleges that currently have undergraduate degree programs in these and related areas.

Undergraduate professional education in agriculture and related areas can offer content, context, and practice for undergraduate liberal study. Higher education in agriculture can also be a model for other higher education as a result of its specified expectations of students and curriculum, its interest in improved teaching and advising, its efforts to construct an intellectual content for learning and for the application of knowledge, and its integration of undergraduates into a professional and disciplinary environment. But the purposes and functions of the agricultural, food, and natural resource systems, which are driving the changes in educational requirements, are changing at revolutionary speeds. The issues of today are created by developments in international trade, health care, and environmental protection, all of which resonate in agriculture, food systems, and the natural resources. As a result, the scope of colleges of agriculture and natural resources is seen as going beyond their traditional disciplinary components, that is, beyond animal sciences, plant and soil sciences, agricultural economics, teacher education, forestry, and wildlife ecology and management.

The purpose of higher education in agriculture will be to provide for the needs of society and industry in a changing world: graduates with flexibility, diversity, perspective, and values. The students needed are those most likely to think globally, to act creatively, to value diversity, to behave responsibly, to respond flexibly in agriculture's new environment, and to interact cooperatively in college and upon graduation.

In defining their priority programs, colleges of agriculture and food, natural resources, and/or life sciences distinguish themselves from other colleges at a university. Education of undergraduates is but one of the roles of many such colleges. Research, service, and extension join education as justification for agricultural faculties. The model is an integrated system of education and scholarship that embodies the basic disciplines, their application, and the markets and consumers of knowledge. Colleges of agriculture in this model are critical elements in the transformation of knowledge for the benefit of society, but in this model, undergraduate education now requires renewed emphasis and new definition. Society demands this. A multiplicity of ideas has been offered for how to reconstruct higher education in agriculture.

The overriding need for the connecting and integrative point of view will require a major transformation of academic thinking. Faculty members must now turn outward from their professions and the institution. Recent enrollments in colleges of agriculture have shifted to biochemistry and genetics, agricultural business and management, nutrition, and natural resources, thus rearranging the landscape of the college of agriculture. There is also a focus on cultural diversity. Both scientific evidence and human values are seen as components.

Students increasingly come to colleges of agriculture without a sense of context. Thus, a sense of context should be provided, but it should be one that is oriented to the future, not one ensnared by the visions of the past. The missions of colleges of agriculture in the future only *appear* to be similar to the missions of the past, that is, to educate professionals for the food systems, the natural resources, and the sciences. The scope and context of the education will be very different from those of the past. Many models and ideas will be injected into the system, and some will be tested. The resultant change will likely be a systemic, that is, fundamental, change. In fact, the constants in the future will be change and diversity.

Managing change and capitalizing on diversity are keys to a sustainable and thriving higher education in agriculture and related areas. Keys to the management of change will come from efforts to understand the system now in place and efforts to develop the principles and concepts of what changes will occur in society and what changes should be undertaken in higher education to accommodate the models that will be conceived in the future. This is the purpose of this study.

In this fashion, we trust that we can take another step from a mere assemblage of thoughts about higher education in agriculture and related areas toward the reality of the future.

1

Higher Education in Agriculture: The Setting and the Need for Change

Donald Vietor, H. H. John, and P. B. Thompson

Key Concepts

1. The content and structure of the undergraduate educational programs in colleges of agriculture, food, and natural resources have their roots in traditional programs, largely in place during the first two-thirds of the twentieth century.
2. The key requirement of undergraduate education in the twenty-first century is that it connect with the national and the global society.
3. The principal forces in the developments of curricula in recent decades have been the growth in scientific knowledge, the changes in communication, and the core and general education requirements of the parent university. However, societal problems span a broader need.
4. Colleges of agriculture have responded to needs in education that go beyond the production components. Business aspects and the life sciences have flourished in some colleges. Environmental sciences are emergent. But insights are needed beyond these incremental changes: the impacts on human health, the understanding of sustainability, the scientific method of problem solving and its limitations, and the public and political perceptions of agriculture need consideration in courses and curricula.
5. The impetus for change comes from both within and outside the college. The elements are the changes in vision, values, faculty, students, and clientele.
6. Systemic, that is, fundamental, change is needed.

Introduction

The land grant college began its mission of enhancing society by educating an uneducated populace that was, at the same time, the principal resource of the society and its principal challenge (Gordon, 1992). From that beginning, novel in its form and noble in its purpose, the success of the land grant system of higher education has had no equal.[1] Yet, as the twenty-first century approached, there was increasing uneasiness regarding the connectivity of the land grant university and its kindred institutions to a changing society (Maw, 1994).

The decade of the 1980s witnessed a sense of disarray, confusion, self-examination, severe criticism from within and without, dedicated planning efforts, and change in higher education in the United States. During the mid-1980s, at least three national studies were commissioned on higher education. A number of regional conferences and individual university efforts were targeted at bringing about change in higher education. The major discussions of this period focused upon topics such as redefinition of liberal education, the requirements for general education and core curricula, the disjunctivness of the undergraduate curriculum, and educational needs of students in the twenty-first century.[2] By the late 1980s most colleges and universities had instituted general (liberal) education requirements and core curricula that were applicable to all students.

Not until the end of the 1980s and the beginning of the next decade did colleges of agriculture begin to seek answers to the same questions confronting higher education generally in the preceding decade. Questions also were being asked about the relevancy, necessity, and survival of the college of agriculture. Agriculture had difficulty in moving from its traditional role and relating to societal changes, to consumer demands and interests, and to the interdependencies of regional, national, and international communities (Kirkendall, 1986). Particularly, colleges of agriculture were slow to accept the public's concern and demand for, say, environmental protection.

Such unease, which is also prominently displayed among traditional agricultural constituent groups, demands from heretofore unserved members of our populace, calls from the university and college leadership itself for change, and beckons for models that confirm the connectivity between the public higher education system and the society it serves (Maw, 1994).

However, much has changed for colleges of agriculture during the latter half of the decade of the 1990s, that is, since the publication of the first edition of this book. Colleges of agriculture, food, and natural resources have stabilized their positions in universities. Survival does not seem to be an issue. The effort to divest colleges of agriculture of their life science components has waned. The opportunities are evident to make larger contributions to university programs and to increase efficiency. However, an examination of the current social and economic environment is a necessary consideration in proposing or evaluating any basis for change in higher education in agriculture.

Historical Analyses of Programs in Agriculture

Kellogg and Knapp (1966) studied the undergraduate programs in the United States colleges of agriculture after World War II, and the study created the first responses to the changing environment of colleges of agriculture. These authors attempted to look to the future at the time of the first postwar decline in enrollment in colleges of agriculture, and an urgency was sensed as to the need to recover their relative strengths in the university. (The second decline occurred in the 1980s after a substantial increase in undergraduate enrollments by an influx of urban students and women in the 1970s. Women continue to increase in proportion. The rural contingent continues to diminish.)

During the early 1960s, the academic programs in agriculture in the land grant universities had many aspects in common. As the missions of the colleges were similar, the general curricular structures were much the same. As Kellogg and Knapp (1966) reported, the average distribution of requirements in colleges of agriculture were as follows:

	Percent
Natural sciences and mathematics	27
Humanities, composition, and speech	10
Social science	12
Agriculture	30
Electives	16
Other requirements such as physical education, ROTC	5

Changes, however, were beginning to occur. Colleges of agriculture (and their parent universities) were beginning to increase their general education requirements. Colleges of agriculture had begun to reduce the number of "technical, how-to-do-it" agricultural courses that were offered. Courses in the upper divisions were consolidated to stress principles. "Round-robin" courses in the several traditional agricultural subject matters were reduced for all majors, and colleges were beginning to reduce the number of specialized curricula.

In the early 1960s, however, importance was given to flexibility in the undergraduate program. The faculty counselor and the student worked toward individualized programs. The programs, however, were faulted in their lack of achievement of high standards that could "attract first-class students."

The goal of the undergraduate degree in the 1960s was largely professional development of the agricultural specialist, which was defined by Kellogg and Knapp (1966) to mean "the managers, technical consultants and advisers, scientists, engineers, educators, and other professionals who manage and lend efficiency to the public and private enterprises engaged in agriculture and renewable-resource developments." The practical knowledge of the time was regarded as ephemeral and Kellogg and Knapp suggested that such knowledge is best gained in the corporate and agency classroom and in summer experiential settings. The undergraduate's education was considered to be best directed toward guided practices in the habits and discipline of learning that may be turned to the "self-discipline of observation, study, and productive reflection." Kellogg and Knapp recommended a specialized education, instead of studying something about each area of agriculture. This was thought to provide for development of an awareness of how one factor in a complete agricultural situation influences the other factors in agriculture. Exploration in depth provided the experience of scholarship. The emphasis on specialists, however, was not to be construed to mean a narrow education. A liberal education included language, mathematics, logic, science, and the humanities and skills in their use. Ability to communicate, which required understanding the way others think, was included as a goal.

Thus, an "ideal" curriculum of the 1960s was defined: (1) a general education that included the basic sciences, mathematics, and humanities, mainly in the first two years, (2) a limited introduction to the broad field of agriculture to sustain the student's motivation during the first two years, and (3) concentration in a major field of agricultural science and technology, along with supporting courses for depth and breadth. The faculties then, as now, were urged to stress the need for skills of reading, writing, and speaking.

Some aspects carried over from such past experience. Some faculty members began to consider the following to be "good" characteristics of contemporary curricula:

1. The program is rigorous. Graduates are expected to be "well educated" and competitive. However, the rigor of the agricultural program in many settings today appears to be measured by the mathematics, chemistry, basic biology including biochemistry, and perhaps the applied environmental science content of the curriculum, but not the rigor in agricultural courses.
2. The program is attractive. Standards have evolved in some colleges to be high enough to attract the better students, and employment opportunities are opened by the selection of appropriate courses.

Alternatively, other institutions may tend to reduce the requirement of the more "rigorous" courses, because they represent "stumbling blocks," in order to attract students into the curriculum.

3. Our programs emphasize the ability to communicate. Surveys of graduates and employers place high values on communication skills, computer skills, and personal characteristics (Litzenberg and Schneider, 1987). A common theme of curricula today is writing across the curriculum and oral and written class reports.
4. Our graduates are employable. Curricula have been often captured by the job, the agency or firm that provides the job, or the professional standards set for jobs that require credentials. This has probably been more prevalent in the natural resources and in human nutrition (dietetics) than in other areas in collegiate education in agriculture.

Whether these are the ideal characteristics to be achieved at any institution is uncertain. But the logic of Kellogg and Knapp (1966) regarding the agricultural curriculum was widely debated and then accepted by faculties in agriculture. It remains an influence on programs of higher education today.

In the 1980s and 1990s, however, differences in curricula and emphases among colleges of agriculture have become greater than the similarities, at least in numbers. Many students in colleges of agriculture have shifted to biochemistry and genetics, agricultural business and management, nutrition, and natural resources. Newer considerations in curricular content include the need for economics and a greater management emphasis. The need for preparation for the international work world is a sensitivity evolving from industrial clientele. Institutions, feeling the pressure of critique, are talking about sustainability, environmental emphasis, and stewardship (Miller, 1995). Ethics and policy are becoming elements suggested by agricultural, scientific societies (Hartel et al., 1994). However, how each of these elements is addressed rests on the initiative and curriculum design within the individual institution.

Unlike the colleges of agriculture of the 1960s, when Kellogg and Knapp (1966) expected that most students in colleges of agriculture would be male and be coming from rural areas, the contemporary college of agriculture student body is increasingly from urban areas, increasingly female, and increasingly diverse in race and ethnicity. The traditional social and ethical context of agriculture—the special relations with the natural world and the values of labor, the community, and rural life—can no longer form the guiding bases of programmatic development in colleges of agriculture (NRC, 1992). Doering (1992) argues that the industrial sector has taken over agriculture and has permanently altered the context for higher education in agriculture.

The Current Environment

The system of higher education in agriculture in the United States will be faced with challenges such as it has not faced within the memory of most of us (Schuh, 1993). Society is undergoing massive domestic change. Within the lifetimes of the students in agriculture, billions of people will be added to the global population. The food system finds itself in a world undergoing a phenomenal reconfiguration of economic and political power. In many parts of the United States, the agricultural land is being lost to suburbanization, the need for stewardship of the ecosystem is increasing, and competition for the resources of the food system is expanding. The public is increasingly concerned with health-promoting and safety aspects of their food.

The changing environment reflects a changing rural America (Pulver, 1994). Rural America is no longer primarily dependent upon farming. Less than one in five nonmetropolitan counties in the United States is primarily farm dependent, although some specific communities may be almost totally reliant on income from farmers. More rural residents are dependent on employment in manufacturing than are dependent on farming. Service-production industries such as health care institutions, computer-software producers, engineering firms, recreation- and retirement-oriented businesses are predominant rural industries. The improvement of rural roads has enabled the patterns of retail trade to move to urban malls and urban services.

The plurality of world views[3] in the current environment, expressed at levels ranging from the local grocery store and community to policy-level decisions in state and national government, is relatively uninformed about the food industry. As the number of people involved in production agriculture has declined, representation in the national policy-making and budgetary decisions has declined as well (Nipp, 1988). A relatively small number of units (less than 280,000) now produce almost 80 percent of food in the United States (Offutt, 1993).

Consistent with national trends, environmental and natural resources, food safety, human nutrition and health, leadership development, and life quality are now included among research programs and in strategic plans of national and state research agencies relevant to agriculture. Consumers, particularly suburbanites, are now identified as the ultimate customers for products of agricultural research and extension educational programs (Bullock et al., 1993; Stauber, 1993). Even as the research agendas of colleges of agriculture continued to evolve, critics have questioned the ability of the agricultural research system to meet the future needs of society (Heichel, 1990). The public agricultural research system is accused of serving peer scientists rather than the larger public community.

One National Research Council statement defines the situation and challenge for education reform in land grant colleges of agriculture as follows: "The land grant system must renew its social contract." The system must respond to a changing student body, a changing clientele, and changing technology for information delivery. Old structures and cultural norms no longer meet fully the demands of these new realities (Bullock et al., 1993). Unfortunately, challenges for renewal and reform are levied in an environment of diminishing resources. In many states, funding for higher education has decreased. Short-term problems that have immediate economic impact compete with education for resources allocated through state and federal governments. Declining enrollments in the 1980s provided justification for redirection of resources from production-oriented disciplines to departments and colleges that offer curricula and expertise for addressing broader concerns. Increasing enrollments through the 1990s have apparently stabilized the trend (Food and Agriculture Education Information System [FAEIS], 1999).

Historically, the rapid growth of knowledge and the needs of special-interest or clientele groups in agriculture have contributed to specialization and the organization of information and people into academic disciplines. Curricula have been developed in terms of essential knowledge and skills within disciplines and of university core requirements in communication, mathematics, hard and soft sciences, and humanities.

Faculty expect students to recall, comprehend, and deductively apply the ever-expanding body of disciplinary knowledge in the context of agricultural problems (Cardwell, 1985; Pennock and Scanlon, 1985). The trend toward basic, reductionist research (Busch and Lacy, 1983; Madden, 1986) has further narrowed the focus of disciplinary courses to new scientific knowledge and away from practice (Esmay, 1991; Kunkel, 1997). This narrowing focus of research and education was used by both agricultural practitioners and the broader public to justify concerns that agricultural research and education are serving professional peers and disciplines more than students and the public welfare (Bradshaw and Marquart, 1990; Heichel, 1990).

The college of agriculture at the turn of the century is often a technical community of scholars with narrowly defined views. Even when problem-solving and management exercises have been assigned to students to encourage integration of knowledge and application of principles (Vietor and Lucey, 1978), the dominant goal has been increased agricultural productivity. It is not surprising that educators are now challenged to include analyses of trade-offs between increased productivity and the externalities of environmental degradation, animal welfare, health and safety of agricultural workers and consumers, and declines in agricultural communities.

Present in each of these arguments for change is not only the underlying theme of the necessity to engage in processes that would reconnect our institutions of higher education to the diverse needs of the increasingly diverse publics, but also the proposition that the connection be made effective, efficient, and sustainable. There is need for further structural changes in the institutional fabric itself, as well as for the establishment (or strengthening) of collaborative efforts with other institutions and governments (Maw, 1994). To accomplish these changes requires a modification and refocusing of the institutional mission.

Encouragingly, colleges of agriculture are examining their programs, engaging new clienteles, and setting new directions. Some colleges have built diversity and intellectual strength and have the ability to meet the challenges of the future. Colleges increasingly include a response to societal issues. Most of our colleges are embracing true strength in science in some areas. Suburbanizing environments in many of these states and competition for resources to sustain viable existing industries, to accommodate development of new industries, and to maintain and enhance the quality of life of the citizenry require nontraditional responses of universities whose programs were historically focused toward agriculture. Accelerated population growth and shifts, a concurrent growth of agricultural diversity, the globalization of education and research, an overall decline of land devoted to agriculture due to suburbanization, changes in markets and market strategies, and an erosion of the environment have been among the factors that have prompted our institutions to engage in futuring/strategic planning efforts (Maw, 1994) as early as during the decade of the 1970s and continuing since.

It can be argued that institutions of higher education in agriculture, particularly the land grant colleges of agriculture, are well positioned to respond to the needs of a United States agricultural and natural resource system that is much greater than the production component. Academic disciplines within agricultural colleges already offer courses about resource conservation; postharvest processing, manufacturing, distribution, quality, and marketing of agricultural products; human health and nutrition; leadership and community development; and ethics. In addition, curricula have evolved in a context where research, extension, and education can potentially interact (Vietor and Cralle, 1992). Simultaneous commitments to quality and excellence in teaching, research, and public service can contribute to a fusion of science and technology, predicated on human needs, that the general science community would do well to emulate (Ruttan, 1991).

But critics say that some colleges of agriculture have allowed the world to bypass them, in part by failing to respond to the significant societal changes relative to land and the food system (Council of Administrative Heads of

to food products from genetically modified organisms and hormone-treated animals, and (4) increased concern about the impacts of agriculture on the environment, food safety, and endangered species. The clientele of colleges of agriculture, particularly those from representative potential employers of graduates, can be of great assistance in defining and achieving our desired educational outcomes. Employers and commodity groups will generally take a self-centered view of educational outcomes but at the same time can provide educational perspectives that are deficient among the faculty. They also have an empirical/experiential knowledge base regarding the operation, management, processing, and utilization of agricultural and natural resource bases that does not exist among faculty.

Colleges also need to pursue aggressively the expansion of their political and clientele bases. As the vision and values of a college change, so must also its support groups. This may mean the forging of new alliances with the broad sector of environmental, animal welfare, organic farming, and similar organizations. The difficulty is not in obtaining assistance from these groups. Rather, it is in developing the process by which such knowledge as is needed can be incorporated into the program in an efficient manner that enhances the educational outcome.

Conclusion

The external and internal environments of higher education in agriculture, food, and natural resources set a need for fundamental change. Externally, there are new publics: new producers, new consumers, new interest groups, and new employers of our graduates. There are expanded issues: environmental degradation, animal welfare, human health and safety of workers and consumers, and the consolidation and growth of industrial agriculture. These are powerful forces at work.

Internally, colleges of agriculture are receiving new students, many of whom have little knowledge of the food and natural resource systems. New faculty, often recruited to compete in a research-oriented atmosphere, also have limited experience in these systems. But, the students and faculty, along with interested constituents, form the human resources for change.

The clear admonition is that any model of higher education in agriculture should again call for reestablishing the connectivity between the higher education system and the society it serves. It can be done but likely will first require reexamination of the values within the institution. Encouragingly, some colleges are examining their programs and are setting new directions.

But most undergraduate programs continue to follow a form fixed decades ago, and the values of the faculty—some changing as a result of restructuring of the faculty of the modern college of agriculture—may not support an integrated education for students in agriculture. It is at the undergraduate level that higher education in agriculture may be the most vulnerable.

The need for change is upon higher education in agriculture whether it is ready or not. While the environment of North American colleges of agriculture has evolved, the values and visions of many colleges may be failing to keep pace or, perhaps more important, to realize the lag in needed fundamental change. The political and societal environment of higher education, the needs of the constituencies, and the knowledge base of education have been transformed during recent decades. The value system in institutions of higher education, with their loyalty to certain elements of the agricultural system and the apparent emphasis on reductive research rather than on undergraduate education, may not be the values needed in the decades ahead.

A note of caution is in order. The societal trends affecting the environment of higher education in agriculture are real, but imperfectly understood. Our perceptions of them may take new dimensions in the early decades of the twenty-first century. But, as Drucker (1994) writes, society will not have the opportunity to resolve the new and looming problems of tomorrow unless the challenges posed by the developments here today are addressed.

Notes

1. Although a substantial portion of higher education in agriculture is carried out in institutions outside the land grant fold, the land grant college of agriculture has served as the model for development of the system of higher education in agriculture in the United States.

2. General education shares many of the goals of modern liberal education, but they are not the same. The goal of general education is to equip students with the skills that they need for individual and social actions with the problems of the present and the future. Miller (1995) suggests that liberal education builds on the liberal arts and assumes that knowledge is valuable in its own right. It is concerned with the preservation of universal truths and the development of the intellect.

3. The way people think about agriculture may be embodied in their *world views.* World views consist of experiences, feelings, emotions, attitudes, values, beliefs, morals, tastes, intelligence, and knowledge expressed in the meanings given to situations and the improvements are preferred.

2

Thinking through Higher Education in Agriculture: Content, Values, and Purpose

Paul B. Thompson and H. O. Kunkel

Key Concepts

1. The knowledge base of the programs in colleges of food and agriculture, natural resources, and life sciences has grown spectacularly, extending far beyond the base of the traditional production technologies and substantially expanding the range of competencies of the faculty.
2. As the knowledge base has expanded, the scopes of education and research of the colleges of agriculture have expanded and have become more diverse.
3. Value connections of colleges of agriculture are being transformed. As the demographic bases of faculty change and the needs for education programs in production decline, explicit and cultural values within the college change. Agricultural colleges share values implicit in universities that emphasize research and external funding, that is, the importance of science.
4. The general change in emphases in colleges of agriculture and their future call for a broad definition of such colleges: colleges of agriculture and their related and modern derivatives are the academic structure that provides the educational, scientific, and scholarly framework for the understanding, development, management, and use of biologically and ecologically based systems and the relevant human resource systems for the benefit of human and natural societies.

Changing Knowledge Base

To address the challenges facing education today, we first have to think through education—its content, its values, its purpose. This chapter is written to do that.

Agricultural knowledge is evolving. It is now and has been a combination of different forms of knowledge—indigenous or traditional, scientific, and industrial—that evolved over time. In the process, radical breaks with the past have not occurred, but the hierarchy of these forms has undergone progressive modification (Byé and Fonte, 1993) and increasing complexity.

Indigenous knowledge of a country is knowledge developed and perfected without particular reference to progress in the development of scientific knowledge. It is knowledge resulting from observation and experience over many years. It remains an important base for courses in international agriculture.

Scientific knowledge is based on scientific principles. The emphases are upon expected effects of a method or technical object. Scientific knowledge is based upon important discoveries in a science such as physiology, nutrition, genetics, molecular biology, chemistry, embryology, econometrics, sociology, and anthropology. Science provides the increasing ability to intervene at different hierarchical levels in agriculture.

Industrial knowledge is knowledge of the work process rather than of the agricultural production process itself (Goodman et al., 1987). It is technological knowledge, that is, science-based knowledge interpreted in relation to changes in the functions of agriculture in industrialized economies. It is also organizational knowledge of simplification, specialization, and segmentation. Industrialization of agriculture is being fueled by a complex of changes in consumer demands, new production and information technologies, goals for efficiency, concerns about risk management, and increased financing.

Throughout technical change in agriculture and food, the agricultural practices that have persisted have been progressively integrated with industrial and scientific knowledge. The extent to which this integration has occurred varies with the economy and the commodity. In much of the world, the convergence of experiences and the improvement of production methods continue to dominate animal husbandry techniques, vine and fruit growing, and the transformation of many food products, even into the twenty-first century (Byé and Fonte, 1993). Both traditional knowledge born of economic practices and trade in food and agriculture and the biological foundations of agricultural production continue as central features of technical evolution in many countries despite the recent decades of industrialization.

Prior to World War II many agricultural practices developed without particular reference to progress in the development of scientific knowledge.

These same practices, closely linked to the biological nature of agriculture, formed the bases on which industrial techniques developed. As these practices evolve, it becomes easier for industrial techniques to adapt to and enhance the diversity of agrosystems. Thus, a large-scale productivist logic of industrial origin has been adapted to the reproductive logic of the biological world. An industrial agriculture was created that utilized business and industrial techniques, that depended on inputs from other industrial sources, and that adopted scientific-based techniques and specialization in its operations. This transformation, too, has been the inherent basis of higher education in agriculture since World War II.

There is a profound difference between productivist orientation of multinational specialized agriculture and the empirical and traditional approach that still characterizes many sectors of agriculture in the developing world (Byé and Fonte, 1993). In the case of industrial agriculture, the production process is segmented into simplified operations for purposes of control. The science is generally clear. In the empirical approach, a group of complex operations must be mastered. The science of the complex is not always clear, although there is an internal logic that must be respected. Thus, the two types of production logic oppose each other: the one is reductionist and the other is complex and requires integration.

The industrial approach has become increasingly dominant in United States colleges of agriculture. It also has left education for the agriculture/food system with discontinuities. And out of these discontinuities, agriculture and its education/research system have become objects of criticism. The environmental and social impacts of industrial agriculture are being questioned. Concern is being raised about the multinational consolidation in the food and agriculture system. Concern is being raised for the sustainability of agriculture, its relationship to food and matters of health, and the economic risks that producers take. And, in truth, animal nutrition is most often taught with little reference to the production ecosystem or to human nutrition. Agricultural marketing may be taught without reference to human consumption. Plant breeding may or may not be taught with reference to food or health or nutrition or problems of intellectual property. Soils may be taught with little reference to the handling of animal waste.

This lack of connection among courses is not peculiar to agricultural curricula. All major national reports on higher education have identified the incoherent curriculum as a part of the problem in higher education. The American Association of Colleges report *Refining the Meaning and Purpose of the Baccalaureate Degree* found that universities are more confident about the length of the curriculum than about its content and purpose (Brentlinger, 1986). The report estimates that one-half of our college undergraduate education falls in the category of training, where training is defined as the study of problems whose solutions are already known, while education is the study of problems whose solutions are unknown.

Farmer (1988) raises the principal issue: "Education in most colleges and universities is fragmented. Students experience the curriculum as a collection of courses rather than an integrated plan of learning. This encourages students to compartmentalize their learning rather than to make connections. Higher education today graduates students with discipline-based minds who will need to function in an increasingly complex and interdisciplinary environment in the twenty-first century." Few undergraduates have been taught to make connections.

The discontinuities in education have a second important impact on higher education in agriculture. The need for a graduate to have a breadth of technical knowledge seems diminished. Specialization, knowledge in depth, may be a preferred basis for employment that, today, seems also to emphasize the ability to communicate technical knowledge and problem-solving competencies. But it, in itself, may not serve the needs for life-long employment. The university impulse that liberal education is central to living "rightly and well in a free society" (Wingspread Group on Higher Education, 1993) may then fill the agricultural curriculum with liberal and general education requirements, while colleges of agriculture wish to turn to emphases of critical thinking, communication skills, and cultural and global perspectives to provide the needed flexibility.

The professional world that current students will face will be vastly different from that of today; these students will be working almost totally in the twenty-first century. What is the knowledge base that will meet the intellectual and social needs in the twenty-first century?

The technical knowledge sources needed for agriculture in the future will be twofold: (1) the knowledge that aims at creating the instruments and practices that can handle the complexity of living systems more effectively and (2) the knowledge base that aims at linking and integrating new techniques (Byé and Fonte, 1993). The first tends to redefine the knowledge base to create new methods. The second goes to the task of improving industrial techniques. They do not answer the same social goals and realities, but both have been involved in the redefinition of the vision and methods of higher education in agriculture.

The New Biology

Observed trends suggest that the food and agriculture system will become more sensitive biologically and managerially. Biological knowledge is undergoing a rapid evolution. Medical education is significantly revised as a result. The new biology has also transformed graduate agricultural education.

A bioindustry has developed upstream from the agricultural and food industries. It differs from agricultural and food industries in its financial origins and in its aims. However, the history of such integration is not new. With the discovery of vitamins, essential minerals, and substances that produced what was then greater than normal growth (antimicrobials and growth stimulants) in the late 1940s and the 1950s, the livestock and poultry industries were transformed from the farmyard to intensive systems of production. The creation of corn and sorghum hybrids and of new seeds meant that the farmer was progressively stripped (freed?) of his control of the genetic heritage of his crops. The new techniques of genetic modification fit the same logic and have the same sort of consequence, which is to attempt to free agriculture from "natural" conditions and set more effective conditions for the industrial production process. This will continue to lead in the search for the ability to dominate nature in the quest for production and productivity, a process that may not be easily accepted by an urban public and that may risk a retreat from industrial trends in the future.

Knowledge derived from science and technology is also incorporated into industrial products (Byé and Fonte, 1993). For example, chemical products often integrate various forms of knowledge related to pest management, weed control, and plant pathology packaged in mechanical techniques to facilitate their use. New knowledge in biochemistry and molecular biology is used to incorporate disease and insect resistance and production factors in industrially produced seeds and seed livestock. Knowledge related to the management of diversity is integrated into computer programs and models. Different forms of knowledge are mobilized to articulate and reinforce industrial techniques rather than create alternatives. In that context also, nature is dominated instead of being used dynamically. Education in some areas will be less important as new knowledge is technologically fixed in the inputs.

Thus, abstract knowledge is appropriated by experts (scientists) and codified, and traditional knowledge loses reference to the agricultural ecosystem (Byé and Fonte, 1993). The systemic unit of animal or plant agriculture is broken up into a group of simple parts to create a constructed system. The people involved in the conception, elaboration, and development of techniques are changing. What may occur is a forced expansion of the knowledge base to require integration with the social sciences.

Changing Natural Resource Sciences

In higher education in the natural resources, the science of the complex is relevant. Forestry education, as early as the 1960s, found it necessary to move away from the primarily industrial management and production emphasis and to give much greater emphasis to the noncommodity aspects of forestry. The education began to give increased emphasis to topics of recreation, wilderness, aesthetics, and environmental quality. The concepts of multiple use and sustainability of production were expanded to include nonconsumptive, along with the traditional consumptive, uses. Although these changes were initially in response to consumer and user demands for a broader definition of forest resources and their uses, they became integrated into the philosophy of the forestry profession and forest education. Though this initially was aimed at the public-owned forest resources, it has become an important part of the industrial and private forest resource education.

Similarly, the wildlife/fisheries discipline has changed its educational emphasis from predominantly production and management of game species to encompass production and management of nongame species and entire habitats. Thus, in the more developed world at least, there has been growing recognition that the information (nonconsumptive) value of wildlife/fisheries resources can exceed their material (consumptive) value. This drastically expands the knowledge base with which educators must deal—from only the mechanics and economics of production to the socioeconomics, ethics, philosophy, and politics of resource allocation and conservation. The conservation issue is especially complex because of the new tools (and risks) offered by advances in genetics and molecular biology.

But coming out of the knowledge base of natural resources is one of the significant new industries for the twenty-first century society—fish farming (Drucker, 1999). The other is biotechnology, and both are logically in the scope of higher education in agriculture. Within the next half-century, fish farming may change the world from marine "hunters and gatherers" to what is analogous to agriculturists. For example, twenty years ago salmon was a delicacy. Today, it is a commodity appearing widely on menus and on meat-market shelves. Fish farming has revised the availability of salmon.

In the case of marine fishery resources, the problems of management are even more slippery because both the developing and developed countries are heavily dependent on marine fish as a source of dietary protein for humans and their livestock, making for industrial overfishing and political tension on a global scale, and because the ecological damage that goes on beneath the sea surface is harder to see than, say, the burning of tropical forests or the draining of temperate marshes. Increasing recognition that wildlife/fisheries share with agriculture a common set of biological and ecological principles and that, indeed, both comprise one resource system clearly points to the need for an interdisciplinary team effort that can be mounted effectively only by embracing systems approaches to education, communication, management, and research.

Evolving Social Sciences

With the industrialization of agriculture, the social sciences have taken on a higher priority. The social sciences focus on various dimensions of human behavior and on the social systems of agriculture and rural America. They have important roles in understanding and explaining factors affecting the well-being of rural people and in providing information for improved private and public decision making.

The dramatic restructuring of society in recent decades, including the industrialization of agriculture, has raised a new set of increasingly complex questions about our society and the economy. As a result, these disciplines are focusing on a broader set of issues than in the past, reflecting the extensive linkages that agriculture has with the rest of the world (Barry et al., 1994).

Agricultural Economics

Agricultural economics focuses on the economic behavior of the producers and consumers of agricultural products, domestic and international markets, rural communities, natural resources and environmental and related public policies. Most of the early focus of agricultural economics was on farm management and marketing. Later, during the Depression, government intervention in agriculture and agriculture policy became increasingly important. Most recently, natural resource, environmental, food, and rural development policies have taken on greater importance. This growing diversity of subject matter reflects the decline in the relative importance of the farming sector and the emergence of new issues and problems.

Undergraduate education in agricultural economics is placing increased emphasis on agribusiness. Justification for this changing curricular emphasis is that the agribusiness is quantitatively different from other economic sectors and that training should include these institutional peculiarities (Lindsey and Martin, 1993). These curricula are built on the complementary relationships with basic management, marketing, planning, and finance offered in business colleges but also provide an appreciation for the unique challenges presented in agriculturally related businesses and markets.

Agricultural Communications

Agricultural communications center on human processes and interactions involved in the processing, dissemination, and use of information. This discipline is rooted in the communication sciences and is concerned with intrapersonal, interpersonal, group, and mass communications. Communications are caught up in the new information revolution.

Rural Sociology

Rural sociology draws on general sociological theory and methods to understand the rural experience, thereby linking description with intervention. Conceptually, rural sociology is the study of social organization and social processes that are characteristic of rural people. The traditional emphasis of rural sociology has included solving the practical problems involved in transforming rural society and preserving the qualities of rural society as portrayed by Jeffersonian values. Future directions include increased attention to community development and increased emphasis on human development.

Agricultural Education

The knowledge base in agricultural education subject matter has changed over time. The Smith-Hughes Act of 1917 provided funds and a philosophy to construct an educational program to educate people about farming. Concepts and production practices about crops, animals, sales and service, products and processing, forests, conservation, and natural resources, and mechanics were taught at the secondary level to prepare young people for work in agriculture. Over time, the philosophy of agricultural education developed to include a broader agricultural agenda (National Research Council [NRC], 1988). Today, programs that plan education *about* agriculture are recommended. Consequently, agricultural education subject matter has expanded to include agribusiness marketing and management in a global economy, public policy, environmental and resource management, and nutrition and health. A major part of agricultural education is career awareness and development, experiential learning, and leadership development.

Philosophy

Philosophy, particularly moral philosophy, has become an increasing contributor to the knowledge base of agriculture, food, and natural resources. Human spheres of economic activity, public policy, health care, scientific research, domestic and family life, recreation and play, and cultural and religious practices present ethical challenges in studies in colleges of agriculture. The response to human and animal welfare, commitments to individual and community life, respect for social justice, allegiances to values and norms of agricultural traditions and natural landscapes, redefinition of agrarianism, and concern for the world's human community, now and into the future, are factors increasingly impinging on the knowledge base. Certain new biotechniques have the potential to alter nature radically.

Ethically critical checks on practices in agriculture and natural resources originate in the public domain of political life, not as values shaped by the welfare needs or economic opportunities of the individual human (Donnelley et al., 1994). The dominant modern ethical theories—deontological or rights based, the consequentialist/utilitarian with welfare concerns for the effects or consequences of actions—have emerged from the domains of human political activity. It is this dimension of critical issues that higher education in agriculture will face in the future.

The Future

The knowledge bases of higher education in agriculture, food, and the natural resources will not be limited to the traditional disciplines and subdisciplines or even to the new ones such as ethics. Elements of cultural anthropology, political science, psychology, history, law, communications, and geography, among others, will be increasingly integrated into education in colleges of agriculture. They, too, will be added to the base of knowledge for agriculture and related areas as integral parts, not simply as elements of the general education of students in agriculture.

New relevant disciplines will evolve and will also impact the knowledge base. Issues such as globalization, resource use, and biodiversity will bring knowledge sources such as oceanography, atmospheric science, ecology, geology, and biogeochemistry into the increased interdisciplinarity of the knowledge base of higher education in agriculture.

As the biological sciences evolve, their increasing complication will profoundly impact the graduate education and thereby the undergraduate education in agriculture.

Human systems have transformed nearly every part of the world. That means the reconnection of higher education to the society it serves will require a sense of society and how society works. But all bases of knowledge—agricultural, biotechnological, life sciences, natural resources, and social sciences—have a social and societal impact.

Implications of Changes in Knowledge

The new sciences are likely to be developing not only new knowledge, but the new approaches should bring the economic, social, ecological, agronomic, animal, and biotechnological sciences closer together and these sciences closer to the basic life sciences. These factors and the questions that they raise will form the theory that guides higher education in agriculture in the future. The practical implications of this development may be that both the content of and the human values in the courses must be systemically changed within existing courses, rather than substituting new courses for the old. Thus, elements of human nutrition can be incorporated in courses in animal nutrition, bioremediation can become part of courses in soils, biotechnological systems can become part of courses in animal and plant breeding, and ecological principles can become integrated into practical animal science and agronomy.

Moreover, curricula may be affected. The knowledge bases of agriculture may be compiled into a hierarchy consisting of biological sciences: molecular systems, physiological functions, organisms, and human interests. Within the biological systems and culminating within human and ecosystems are aspects that relate to management, strategies and systems of production, land and water resources, landscapes, human capital and community, markets, and consumer interests (Kunkel and Hagevoort, 1994).

Changing Context

Science-based education in agriculture involves two transformations: a transformation linking knowledge to technologies and techniques and a transformation linking techniques to their social function. Specialists will be needed, to be sure. An educational goal for tomorrow will again be the education of managers of complex biological systems, but the integration will be different from that of the past, where mainly the internal logic of the particular agriculture controlled the logic of the education. The integration will be that of linking technical knowledge and personal development with a societal function.

As an example, in the American context, food is not simply an energy-providing, nutritious product destined to guarantee the low-cost reproduction and maintenance of the work force. Food is involved in more complex functions: health-giving, recreational, social, cultural, ethical, and even religious functions. Food quality cannot be reduced to the chemical, nutritional, toxicological, or physical criteria of the product, nor can food quality be reduced to the criteria applied to industrial and commercial demands. Quality implies, instead, that not only the product must be considered, but also responsibility in the ways in which the food is produced, transformed, and used and judged to be safe. These elements should join nutritional needs and costs in the first-rate higher education in agriculture for the future. Such vision of interrelationships takes on even greater meaning when an international scope is integrated into the thinking. Malnutrition is a health problem throughout the developing world, but the solution of malnutrition is not always found in health delivery. A shift in the price of legumes, for example, may have more effect on a population's nutritional health than a large number of health centers (Berg, 1993). Agricultural education that neglects the global dimensions of food needs is only half-way sufficient.

Certainly the twenty-first century knowledge base of higher education in agriculture will extend beyond the industrial approach. For example, the knowledge base related to food safety—pathogens, carcinogens, toxicants, real and suspected safety factors of genetic modification—will be a part of the course content and curricular material in the future. Impacted directly by factors in food are the economics and politics of international trade. The European Union has banned the importation of "genetically modified organisms," while producers in the United States planted an extensive acreage of genetically modified crops and now face market disruption. Higher education in agriculture will likely include elements that provide a practical knowledge of how to cope with the conflicts of the uncertainties of science and the perceptions in society.

Thus, emphasis on the technological approach may also have impact in the internationalization of education. Science and technology are variably relevant in the agricultures of different countries. Techniques of breeding plants and animals and creation of genetically modified organisms are as important in a developing country as in the United States. But when the technology produces a new crop variant, for example, one that requires seed to be purchased each year rather than saved from the crop, a system of agriculture is affected.

The environmental focus must become much stronger in our colleges of agriculture, to the point where it is seen to be as fundamental to agricultural curricula as is basic biology, mathematics, chemistry, and communication. A theoretical or systematic knowledge of techniques for the survival and sustainability of a cultural and natural ecosystem presupposes an understanding of biological and ecological processes plus a learning system that embraces philosophy or theology to define the intent and purpose of human activity. Then, skills, management, and decision making (with appropriate incentives) can use available wealth and resources to provide the environment desired.

Integrated scholarship should be incorporated in higher education in agriculture. Kunkel and Hagevoort (1994) define integrative scholarship to be concerned with biological, physical, and social issues and values and their interconnectedness in a context. The integrated focus is also more than being multidisciplinary; it is a frame of mind, a thought process. Boyer (1990) and Malone (1994) carry the point further. They speak of the cascade of knowledge that consists of the discovery, integration, dissemination, and application of knowledge. Integration is an important intermediate step in the flow of knowledge that comes from science and observation into use through education and application.

In the real world of tomorrow, if an agricultural program of higher education of any size is to succeed, it must be multidisciplinary, including people and knowledge that understand social science, economics, and now cultural anthropology as well as the disciplines of agricultural sciences, nutrition, and the natural resources. Leadership depends on qualities that are outside any particular discipline. This is the content of education for managers and other professionals who may or may not continue in careers related to agriculture and the natural resources.

Agricultural education should be seen as a continuum. It is a continuum that reaches from basic biology, chemistry, and economics through production, marketing, and human needs and ethics to resource recovery and sustainability. Higher education in agriculture has the obligation to handle issues such as the environment and to make connections with the health fields. Linkages to all such areas must be pervasive. Leaders in agriculture must understand the whole process. If the linkages are not understood, then agriculturalists may well be irrelevant.

Changing Values

Two matters grow out of the values of the land grant system. The "mission" of the program emerges out of traditional values and the heritage of the land grant system. The vision of the program rises out of the beliefs and values of individuals and the particular institution and the opportunities for the future. The vision and the mission are not the same, although they may be derived from common values.

Proposals for extensive reform of agricultural education presuppose a theory of value in a dual sense: in that education is a goal-seeking, or normative, process (hence some notion of value is implicit in the very idea of education) and in that reform implies a modification of whatever goal-seeking educational processes are currently in place (hence some alternative set of values is presumed to be better). This theory of values also requires assumptions about how individuals working in higher education in agriculture can alter the system of education to serve an alternative vision, and in describing constraints to such change. At the least, the theory of values generally implies that the agricultural colleges or universities themselves possess values.

The suggestion that organizations have values is somewhat metaphorical, so clarification is in order. The term "values" is used here to indicate the concepts, rules, and normative beliefs that guide an organization from the inside. Values are internal in the sense that they can be properly attributed to the organization itself, or to its officers and staff, as opposed to external forces and constraints to which the organization must respond. There are at least four distinct ways in which an organization can be said to be guided by its values:

1. Individuals in the organization have values in the common sense of the term. They have beliefs about what is and is not important, and what is and is not proper conduct. If asked, individuals articulate these beliefs explicitly. When enough key people in an organization share these explicit values, it becomes plausible to ascribe the values to the organization itself.

2. Individual behavior is also shaped by tacit assumptions that most people would have great difficulty in stating without substantial coaching. The tacit assumptions reflect cultural values. People come to adopt these values implicitly in their attitudes toward nature and toward other people. If enough individuals in an organization come from a common demographic or cultural stock, it is plausible to ascribe these cultural values to the organization itself. These values are, nevertheless, the values of individual human beings, and if there is enough turnover in faculty, these values will be replaced without significant institutional change.
3. The organization, particularly the educational institution, has an explicit infrastructure that consists of its rules and policies and also has an organizational chart that defines its divisions and departments and stipulates a hierarchy of authority, in ideal terms at least. This infrastructure will be a significant internal determinant of actions and productivity, within an agricultural college, for example, yet it may operate substantially at odds with both explicit and implicit values of the individual in the organization.
4. Organizations also have a host of informal procedures that operate "off the books," and which are not affected in policies or organizational charts. These informal procedures allow individuals to accomplish goals that would be impeded by either a lack of definition or an outright contradiction by the formal structure. These procedures represent a hidden rule book of sorts that allows individuals to act collectively, but which may be very difficult to identify empirically.

How do each of these four ways in which organizations have values influence agricultural education today? It is clear that any given agricultural college can differ from any other one in any or all of the four respects. Nevertheless, there is no doubt that there has been a certain amount of explicit agreement about the purpose and goals of agricultural education in the past. During most of the twentieth century, faculty, administrators, and students shared the assumption that the primary purpose of undergraduate education in the agricultural sciences was to prepare students to operate farms, ranches, and agricultural industries in a competitive environment that has been becoming increasingly complex and technical. However, faculty and administrators shared the explicit belief that education must not be so narrowly tailored to its primary purpose that it would be useless for students who would not return to the farms on which they had been raised, even though, historically, few graduates had done so. As such, higher education in agriculture should, it was believed, be capable of preparing students from farm backgrounds for an array of careers broadly associated with agriculture, including employment in agricultural input firms, in food production and processing, and in those elements of the public sector dedicated to servicing agriculture.

To the extent that these beliefs were characteristic of the values held by agricultural faculty of the past, they committed agricultural colleges to the value of serving a particular constituency of students by preparing them for gainful employment. Although this commitment did not exclude broader educational aims—pursuing education as intrinsically valuable for human fulfillment, for example—it did narrow the focus of agricultural colleges to students who were either coming from or heading to rural communities and who had food and fiber production in mind as their primary reason for being there. Many agricultural colleges added programs in natural resource management that broadened this focus, but this addition did not, in most instances, change the assumption that education was primarily, if not exclusively, a means for securing future employment or livelihood. Agricultural faculties, thus, have had (and to some extent still have) the value that agricultural classes and degree programs are successful when the students enrolled in them are successful at seeking a livelihood in the general area of agriculture and natural resources. But what ought to be changed is that the livelihoods can come from a much larger arena of work that supports human society beyond agriculture and natural resources. The explicit values of agricultural faculty and administrators are extremely important.

Values common to many if not all agricultural colleges are more likely to be found in a second source of values. There has been a remarkable consistency of background and experience among faculty and administrators of agricultural universities, almost all having come from "a farm too small" and having obtained doctorates from other United States agricultural universities. This situation suggested that tacit or cultural values within agricultural colleges would be a major formative influence. The now-retiring group of faculty and administrators tacitly assumed that agriculture was to be understood as food and fiber production. Their commitment to production was ingrained, so much so that when natural resource disciplines began to be incorporated into agricultural programs they were often evaluated in terms of their contribution to production and economic growth. Yet the main thought of many who called on agricultural education to change in the latter decades of the twentieth century has been to suggest that cultural, environmental, or consumer practices are as constitutive of the system of agriculture as is production. The faculty's and traditional clientele's inability to understand these values on a par with production has been a barrier to change.

An important source of values for higher education in agriculture is the formal rules and organizational structure of the agricultural college. Laid out in comparison to one another, the organizational charts of agricultural colleges have had a remarkable similarity, especially when compared with those of other educational institutions. In spite of many variations in precise staffing and reporting details, as well as in size and areas of emphasis, it is doubtful that anyone will mistake the division of departments and administrative hierarchy of an agricultural college for that of another college. Furthermore, the rules for evaluating faculty in agricultural colleges are far more likely to include provisions for rewarding the development of technology than are those in the arts and sciences.

Clearly, the demographic base of agricultural colleges is changing; many more faculty will have urban backgrounds. The rapid embrace of molecular biology has broadened the range of experience and interest among faculty at doctoral-level institutions. All this suggests that both explicit and tacit values of faculty may change in the future. To the extent that molecular biology faculty members have less personal experience and training in the cultural and social dimension of rural life, the result may be a *reduced* emphasis on cultural, environmental, and consumer dimensions of agriculture. If the human health sciences were to serve as the model, this would precipitate a de-emphasis of undergraduate education so dramatic as to make further discussion of reforming the undergraduate curriculum an idle exercise. Similar transformations of the agricultural college to an environmental science institute may have the same effect. All of this serves to emphasize the importance of research as a determinant of organizational structure and, thus, of capabilities for undergraduate education.

Values are complex, but two forces are operating on the value connections of agricultural colleges. One is that, as the demographic bases of faculty change and as the need for education programs to prepare people specifically for a career in agricultural production declines, explicit and cultural values within agricultural colleges will change. The normal process of faculty retirement and replacement will assure this. What is less clear is whether this change can or will be directed toward an alternative vision of the land grant mission, or what people need to know about food, agriculture, and natural resources. The second force is not unique to agriculture but may operate more persuasively within an agricultural college. The organizational values implicit in universities include strategic elements that militate against serious attention to teaching and especially against teaching strategies that require time-consuming individualized instruction, experiential learning, or student evaluation. This problem ought to be viewed as especially serious among faculty committed to a work ethic.

Intellectual Purpose of the College of Agriculture

Today, institutions of higher education in agriculture have different scopes, different aims, and different environments in which they operate. Significantly, they also have different names. A trend that has continued in recent years has been to identify the college as a College of Agriculture and Something Else. The predominant forms of expanded names are College of Agriculture and Food and/or Natural Resources and College of Agricultur(e/al) and Life Sciences. One college of agriculture, Cook College of Rutgers University, has adopted a designation that does not restrict identification with specific disciplines or resources. Within these scopes, various descriptors of the intellectual purposes of the college reflect education, research, and extension in agricultural sciences: food sciences; natural resources; food production, consumption, and protection; environmental science; marine sciences; agricultural industries; life sciences; food, fiber, health, and aesthetic values; the biology/biotechnology, production, marketing, consumer, and recycling continuum; and the environment, agriculture, food, nutrition, health, and medicine continuum (Kinsella, 1993).

A survey among the participants early in this study revealed that there is a diversity of views. This diversity of vision seems evident as well today. Some phrase the intellectual thrusts that the college should have in the framework of both students and processes. The purpose of the agricultural institution is stated as to provide the public and students with the knowledge, skills, and competencies to understand and to manage effectively and efficiently the biological, physical, social, economic, and political dimensions of contemporary and emerging issues related to the ecosystem. Others emphasize that the college should work in conjunction with the entire agribusiness community and agricultural industry in teaching, research, and service activities. These activities should be centered around preparing individuals to understand and respond to the interrelationships between agriculture and society. Other stated intellectual purposes are to develop students' professional and technical knowledge, skills, and abilities related to the field of agriculture in its broadest definition; to develop the abilities of students and other clientele to effectively resolve a broad range of social, economic, political, cultural, and technical problems; and to develop communication skills and abilities, personal values, and interpersonal skills.

Although perceived intellectual purposes of higher education in agriculture are multiple, there is some commonality in the traditional foci of higher education in agriculture: the management of land, water, and vegetation and the manipulation of plants and animals (biological systems) to meet human needs are the traditional foci. This has been largely interpreted to mean management of agricultural and natural resources to meet more efficiently the food and fiber needs of the world. But, in keeping with today's world and the thrust of most colleges of agriculture, the described scope includes fundamental biochemical and ecological sciences; ecosystem management; environmental quality; conservation of biodiversity; sustainability; food safety; interfaces among land, water, and air in the production of consumptive and nonconsumptive products from our resources; scientific certainty and uncertainty; and the interactions of these aspects with human, cultural, and ethical values.

Higher education in agriculture can be defined today to include elements of science; biotechnology; land and water resources; production processes; trade, marketing, and use of agricultural, forest, fishery, and biotechnological products; transformation of food; consumption and human health; recycling; and public policy. Digital technology, often referred to as the information revolution, will amplify these components as it simply allows the routinizing of what is already done (Drucker, 1999).

The intellectual purpose of a college of agriculture must also be set within the context of the parent institution's liberal and professional educational values, goals, and mission. All collegiate units of education in agriculture, food, natural resources, and life sciences do not encompass the entire scope. In fact, individual educational and research units define niches for themselves. But the use of a broad definition of common scopes is not only desirable but may be essential for the undergraduate educational system relevant to agriculture to serve its increasingly diverse clientele.

Smaller colleges of agriculture have many of the same goals as the larger colleges. All wish to recruit exceptional students. The academic approaches could be much the same regardless of the size of the college, that is, an emphasis of science and practicality. Institutions focus on what they can do well and market themselves accordingly. Though colleges may not be able to provide education in the full scope, the education that is offered can be of higher quality and productivity than any other college in the university can provide. Fundamentally, colleges of agriculture should redefine their clientele, regain the confidence of the public, and position themselves to recruit the best students possible.

Colleges of agriculture have the special capability and, thus, obligation to educate the general public about agriculture in the context of human population growth and changing demographics, the needs for food and nutrition, world trade, environmental quality, ecosystem dynamics, and management of renewable natural resources. This is more important than ever before because the public is increasingly aware of the realities of food and fiber production and the interrelationship between cultural influences and ecological systems and has serious misconceptions about these areas, but is increasingly empowered (at least by proxy) to impact the human life-support system. There is a need to develop courses and programs that appeal to nonagricultural areas of the university and that have creditability with students across the campus. Environmental science is one such avenue. The universal need for food and its societal implications is another.

Conclusion: The New Definition

The knowledge base of the sciences in the colleges of agriculture will likely respond in the future more to needs of the consumer and the stewardship of natural resources than to production aspects of agriculture. The knowledge revolution (in contrast to the information revolution) will likely demand reconfiguration of the knowledge base. The three components of agricultural and natural resource knowledge—traditional, scientific, and industrial—will add a fourth component, social concerns and moral studies. But colleges of agriculture are also likely to build on and integrate the "explosion" of biological and ecological knowledge and the growth of economic, sociological, and normative knowledge. We can expect the theoretical underpinnings and value system to take new turns. The challenge to higher education will be to place these aspects in perspective while implementing change.

In the twenty-first century, cultural values within the college of agriculture will change. The historic values of colleges of agriculture are the more likely to be displaced, but it is not clear whether the change will be directed to a different vision of the mission of higher education in agriculture and related areas. The institution will need to be responsive to the changing values and nature of society.

This critique of higher education in agriculture is not intended to disparage what was taught in the past. Higher education in agriculture has had a distinguished history. But there will be new models for the future. A redefinition of the college of agriculture and agricultural sciences as a body of knowledge could permit many models to develop, each of which will find its context.

We propose, as a generic and theoretical form, the following definition: Colleges of agriculture and their related and modern derivatives are the academic structures that provide the educational, scientific, and scholarly framework for the understanding, development, management, and use of biologically and ecologically based systems and the relevant human resource systems for the benefit of human and natural societies. The college looks at the total life-support system.

This theoretical form is an academic view. It is not intended to replace or change the pragmatic missions adopted by colleges of agriculture that include experiment station and extension service organizations or functions. It, however, reflects changes currently occurring in programs of some colleges, and we believe it provides a basis for continued academic vitality. It provides a descriptive framework for envisioning higher education in agriculture and related areas in a rapidly changing society.

3

Envisioning Higher Education in Agriculture

H. H. John, Christine Townsend, and A. Gene Nelson

Key Concepts

1. The vision of how things should be is the beginning point of systemic change in higher education in agriculture. The vision is the launching point in thinking about changes. It is not the same as goals and plans but a definition of how we would like to see things.
2. What flows from the visions includes the desire for excellence, the quest for flexibility in both the curricula and the graduate, and the articulation of goals into actual educational outcomes.
3. Faculty perception of the future graduate as a leader may be the easiest and most effective systemic change that can be made.
4. Student learning is unambiguously linked to effective teaching.
5. Programs should be tailored to meet the needs of the students admitted to the college, not to the convenience of faculty and staff. Teaching and curricula should be designed to invite, encourage, challenge, and support students in their intellectual and ethical development. They should be taught how to think, not what to think.
6. There is a hierarchy in education. Knowledge flows from discovery to integration to dissemination to application. Graduate study emphasizes discovery. Undergraduate education emphasizes dissemination and application. The missing link is integration of the knowledge. What is called for is the integration of the entire land grant system: research, extension, and teaching as well as the knowledge base.

Visioning

The basic premise in this book is that systemic change will be required; mere improvement or incremental change is not likely to meet the needs of the future. What then is systemic change? In part, it may be a change in the campus as we know it. It may be places of education linked electronically rather than limited to a single plot of land. It may be reorganization of the content of curricula and courses or a significant change in intended learning outcomes. It certainly involves the types of faculty that are employed. It is likely that it must be a collective effort of all institutions of higher education in agriculture. It may be a reorganization of the college or administrative structure, departing from the traditional disciplinary model. Underlying any of these aspects of change is the vision of the institution.

The vision is not the same thing as goals or plans. The vision is rather one that defines how we actually prefer to see things. It is a compelling statement of what we want to create. It deals with how we see the situation or institution instead of the strategies set to achieve the vision. It has to do with how we see the landscape. If we see things even in a somewhat different fashion, it will have a tremendous impact and yet may not require changing names of courses or necessarily changing the structure of the curricula. Implementing change does not mean a structural change so much as it means articulation of a different vision for the program in higher education from which structural change may or may not occur.

Being visionary has the quality of sensing the future. The vision of the future has the potential to let one "see" the very broad picture, the place, and the functionality, in this instance, of the agricultural college within it. It provides a view of the "sense of place" for food, agriculture, and the natural resources. Finally, it provides focus to the important differences that occur as a result of systemic change. It is a combination of intuition, good sense, creativity, and courage.

The Vision Statement

Creating a vision of the desired state is the first and, most would agree, fundamental step in initiating change. A vision provides a guide or sense of direction to the change process. It helps all members of the organization to literally visualize the future of the organization. According to one study, "We found no effort to produce strategic change was successful without a new vision" (Benningson and Swartz, 1987). Coming up with an unambiguous definition of vision, however, is not easy.

Richards and Engle (1991) define a vision statement as "a document describing the way things could be. It is a declaration of the organization's most desirable future; it describes the faraway lights, and invites the organization on a quest to reach them." A vision is an attempt to articulate what a desired future for an organization would look like. Vision statements tend to be oriented toward customers, provide focus for employees, describe organizational competencies, and indicate standards of excellence. The most effective vision statements are short and direct. For example, Disney's vision is simply "to make people happy."

A vision statement is not the same as an organization's mission, its strategic objectives, its goals, or its values. The vision statement should, however, be consistent with these elements of the organization's plans, and in fact, it should provide the direction or inspiration for these elements.

In describing the characteristics of good vision statements, Jick (1993) lists the following: clear, concise, and easily understandable; memorable; exciting and inspiring; challenging; excellence centered; stable, but flexible; and implementable and tangible. The experts disagree on who should be responsible for coming up with the vision statement. Richards and Engle (1991) state, "creating a vision statement is the responsibility of the highest level of management in an organization." They argue that the vision statement is best written by one person, that being the senior person in the organization. On the other hand, Collins and Porras (1991) ask, "Is vision setting only for CEOs? We don't think so. Vision setting should take place at all levels of an organization, and each group should set its own vision—consistent, of course, with the over all vision of the [organization]."

Visions can be developed using several different approaches, but the key to success is instilling commitment to the vision. According to Jick (1993), "It's not just having a vision—the 'right' one, of course—that counts, but also one that is well accepted and can be translated into an actual behavior."

Thus, visioning is what is wanted, and values are a part of that. The vision that we seek is what we want to be (the future). The vision of the future goes beyond our current paradigm, our current expectations, and perhaps our current beliefs as to what is important. It invites writing new rules.

Envisioning Excellence and Change

There are three broad questions that should be considered when proceeding in a studied evaluation of our curricula and efforts to achieve excellence in agricultural education: (1) How does one plan for excellence in education? (2) What courses and learning experiences should be included in the curriculum? And (3) how should one teach the courses and structure the learning experiences to meet the criterion of excellence?

It is doubtful that future excellence can be achieved or maintained without systemic evaluation of it in our colleges of agriculture and their curricula. Farmer (1988) notes that a criterion for excellence correlates with the characteristics of students at graduation and not at the time of first enrollment. Admission criteria relate to what a student brings to college while exit criteria are a measure of the contribution made by a college to student higher education. Faculty members have traditionally been active in the establishment of admission criteria with little or no concern about exit criteria. This emphasis needs to shift if excellence is to be accomplished.

Excellence in agricultural education must be looked at from the perspective of educational excellence in general as well as excellence in the disciplinary content (Farmer, 1990). Some of the requirements to achieve excellence are

- congruence between student goals and those of the institution;
- that academic preparation for students or lack of it be honestly assessed and reflected in the curriculum. Faculty need to design and implement a curriculum for what students need to learn rather than what faculty want to teach. Students must accept responsibility for learning what is needed to enter the curriculum;
- creation of a learning environment which encourages students to be committed to learning with the desire to become active, independent learners;
- that faculty members not only keep up in their discipline but actively relate their specialized knowledge to general knowledge outside their disciplines;
- that faculty members demonstrate a commitment to the spirit of liberal education by an openness of mind. Education is a journey, not a destination;
- that faculty be familiar with theories of learning and utilize this knowledge in the instructional design of their courses; and
- that the curriculum result from an integrated academic plan and not just a collection of courses.

Throughout the literature regarding agricultural education we see references to the rapidity of change in all aspects of society that we and our students face.

Writers such as John Naisbitt (1982) and Alvin Toffler (1970) talk about being overwhelmed by change and shocks to our perspectives of the present. A double problem faces us today (Brentlinger, 1986): what is the future going to be and how do we design a curriculum to meet the future? Many of the contributors to *Agriculture and the Undergraduate* (NRC, 1992) call for curricula that contain flexibility, both in terms of course work and in terms of graduates' ability to adapt.

Toffler, in *Future Shock* (1970), states he is appalled by "how little is actually known about adaptivity, either by those who call for and create changes in our society, or by those who supposedly prepare us to cope with these changes."

He continues, stating that those of us within academia who talk bravely about "educating for change" and "preparing people for the future" basically know nothing about it. Toffler adds that for a student to "understand the past or even the present [the student] must learn to anticipate the directions and rate of change—learn to make repeated, probabilistic increasingly long-range assumptions about the future—and so must [his] teachers". In *Megatrends,* Naisbitt (1982) says that the future is simply "moving from the specialist who is soon obsolete to the generalists who can adapt".

Bretlinger (1986) developed a set of criteria, based upon the literature, to judge a curriculum purported to educate for change and thereby prepare students for life in the twenty-first century. Questions that inform the process include

- Do the skills of decision making, problem solving, creating, evaluating, analyzing, and acting take precedence over the acquisition of knowledge?
- Are students educated to contend with problems that do not now exist, learn to deal with the unknown, and be able to structure problems to grasp what the result will be, and will they be equipped for change?
- Is attention given in the curriculum to know how knowledge has been created, to method and styles of inquiry that have led to its creation; in short, is there emphasis on the "what," allowing the culture of the time to decide how one is to achieve it?
- Is there provision for a great amount of individuality within each student's degree program, which may mean some movement away from standardized, rather fragmented series of required subjects?
- Is each student taught how to classify information, evaluate its veracity, change categories when necessary, move from concrete to abstract and back, and synthesize information in realistic ways?
- Will the student have learned how to learn upon graduation, having found somewhere in the experience of education a "life know-how"?
- Is there a recognition of the coming leisure society that may place more emphasis upon interpersonal relations, human development, and the dignity of humankind?

Overall curricular change will be driven at the collegiate level by the faculty as they develop their particular institutional world view and articulate this into the educational outcomes desired. At this level, it is necessary to address some questions about the curriculum, such as

- Should there be a core of knowledge, above and beyond general institutional educational requirements, for all students graduating from a college of agriculture?

- Should such a core be different or in addition to core requirements for a generalist versus a specialist type of career?
- Where do the issues such as biotechnology and the concomitant matters of patents and who owns the intellectual property, ethics and values, environment, and internationalism fit in the curriculum?
- Do interdisciplinary courses have a place in revised curricula? If so, how?
- Can modular courses be of value in providing differing degrees of depth for specific subject matter?
- Should specialization and preparation for graduate studies be the focal point of the curriculum? Or should it be general education? Can it be both without detracting from either?
- Where do professionalism, accreditation, and certification enter?

At the departmental or discipline level, the faculty as a group will have to determine its own particular vision and desired educational outcomes. The foregoing questions are equally applicable to such deliberations.

Envisioning a Prospect: The Graduate as a Leader

Up to this point in this chapter, the emphasis has been on the issues of systemic change of educational outcomes through vision, values, and curricular change. Much can also be done to make significant improvements in the educational outcomes by changes in individual courses and by the use of new and innovative teaching methods. But the way a faculty member sees the incoming student as a future graduate may be the easiest and most effective systemic change that can be made.

As one approach, the institution can accept as a principle the vision that a capacity for leadership in all aspects of work should be a common goal of higher education in agriculture. What sort of educational twist would be required? As an ideal, such a goal would be to prepare students for leadership in society, the agricultural industries, science and education, government, financial institutions, natural resources, or wherever the graduate may find opportunity. Beyond their technical preparation, future graduates of higher education in agriculture should be able to relate to changing issues and forces influencing food and agricultural processes and products and the natural resources. Graduates may be asked to develop creative and proactive initiatives to solve consumer concerns, deliberate ethical issues, mediate interdisciplinary dilemmas, and create effective policy. Cohen (1990) has written, "Leaders are made, not born. If you want to be a leader, you can learn how in the same way that you learned other skills." Undergraduate education in agriculture can take the responsibility to prepare graduates to lead.

Leadership education can prepare students with tools to integrate the technical, economic, social, and ethical spheres of agriculture. The "skills" of leadership can provide graduates with mechanisms to succeed in the leadership of organizations by enhancing their decision-making skills, communication techniques, problem-solving abilities, and team-building expertise. Coupled with technical knowledge, leadership education allows graduates to develop a comprehensive outlook to succeed in diverse agroindustrial experiences.

Leadership education is not a new concept to the food and agricultural industries or colleges of agriculture. Since the early 1900s, leadership concepts were taught at our institutions as a component of agricultural teacher and extension agent undergraduate training. Students in these programs carried their leadership knowledge from the university to high schools and counties. They taught leadership skills to their youthful clientele via the Future Farmers of America (FFA) and 4-H organizations. An even larger vision of leadership capacities of graduates will be needed for the future.

The objectives of leadership education are to provide students with the tools to influence a group to complete a task:

- Leadership education improves the skills and knowledge required by the students to perform their roles and functions within groups. Particular attention is given to the areas in which students experience doubts and uncertainties.
- Leadership education increases students' understanding of individual and social forces that operate within groups and which affect groups from outside. By understanding what motivates people and how they tend to react to different situations, students develop their own skills and confidence in communicating and dealing with people.
- Leadership education increases the morale and self-confidence of students so they think of change and development as opportunity and not as a threat to their position.
- Leadership education is concise, particular, and relevant. It is an ongoing program of training experiences related to personal and professional events rather than a series of infrequent and unrelated courses. To be effective, it provides a regular reinforcement for students.

New Concepts of Instruction

The evidence reviewed by Pascarella and Terenzini (1991) suggests that the real quality in undergraduate education resides more in what institutions do programmatically than in the resources and reputation, that is, student body selectivity, prestige, educational resources, large library, and scholarly faculty.

They suggest that undergraduate educational quality may depend on factors such as curricular experiences and effective general education; patterns of course work; quality of teaching; frequency and focus of student-faculty nonclassroom interaction; the nature of peer group and extracurricular activities; and the extent to which institutional structures and policies facilitate a high degree of student academic and social involvement.

Innovative methods are being developed and can be incorporated into teaching. It is evident that good, motivated teachers, well-equipped classrooms, and a supportive administration are as important as the curriculum. Programmatic development and teaching are two sides of the education model. The need is to develop the learning outcome and teach accordingly.

Student learning is unambiguously linked to effective teaching (Pascarella and Terenzini, 1991). Instructional skills and course structure are particularly important. It is important that ideas of good teaching be made available to anyone who enters the classroom to teach. Successful examples will lend credence to the importance of improvement of teaching. Wide exposure of faculty members to successful examples provides opportunities for faculty members to think about their own courses and how part or all of an idea can be incorporated into a class. Innovative teaching workshops are some of the best venues for such exposure.

New capabilities can also lead to effective change. For example, digital technology and the Internet enable the educational thrust to turn to the ability to emphasize application and synthesis instead of depending on recall and comprehension as the basis for learning.

Institutions can shape the academic environments in ways that invite increased student involvement (Pascarella and Terenzini, 1991). The concept of learning communities of students may be one of the more effective concepts to introduce in the educational process. In such a process, a group of students take courses together and work as a team on an assigned subject. Honors programs are designed in a similar manner. The fact that all students in a team may not "carry the load" is but a reflection of what is often encountered in the "real world." Less emphasis may be placed on the details of what students are learning and more on the quality and integrative nature of learning.

Improvement of teaching can occur through course restructuring with utilization of modules, internships, and enhanced seminars and inclusion of current issues as well as electronic equipment. Innovative teaching requires time, effort, and money. The focus, however, should be on the learning because with the achievement of learning, a certain quality of teaching may be presumed.

In the immediate future, quality of the education of a graduate may be evaluated more on the selection of courses and experiences in the individual students' program than on the grades accumulated, although rigor of scholarship still will be reflected in the grade point ratio. Thus, questions arise: Is there a need for professional licensure and accreditation? Should there be increased emphasis on professionalism? Should soil scientists, for example, complete a fixed body of knowledge and be so certified? The dietetics emphasis in nutritional sciences has such a certification.

Emphasis on cooperative education and internships has had some importance. Such experiential education provides a route into future employment as well as aid in the selection of a career. In addition, and perhaps more important, such experiences provide focused opportunities to both curricular and noncurricular learning within a setting consistent with the students' chosen field of study. Such experiences foster the further development of higher order cognitive skills and allow the students both to demonstrate a wide range of competencies associated with their personal, educational, and professional development and to expand upon these competencies in a work world. But too few students take advantage of such experiences, and other models are needed. The learning community may be an alternative (see chapter 14).

Putting the Student First

It seems axiomatic that colleges of agriculture, with their traditions of interest in student welfare and sense of responsibility to those in agriculture, should tailor their programs—curricula, schedules, support services, office hours—to meet the needs of the students they admit, not the convenience of faculty and staff (Wingspread Group, 1993).

The precollege education and experiences of students in food, agriculture, and the natural resources are largely changed from what they were in past decades. Entering students have less work experience. The students of today are not less capable, but they are different from their predecessors. Thus, questions arise: Should the focus be on preparation for careers or should colleges of agriculture prepare students in a more general education framework? The answers to such a question will be varied, dependent on the value system of the institution.

Clearly the demographic base from which agricultural colleges attract students is changing. Colleges of agriculture must now position themselves to attract and recruit the best students, including Hispanic, African, Native, and Asian Americans. Women now comprise nearly half of the enrollments in agricultural colleges (FAEIS, 1999). Colleges of agriculture will be required to provide a diversity in the academic program to meet different needs created by an expanded range of ages and diversity of the student body. Older students, women and men who have children, students from different cultures, physically handicapped students, students from rural areas and from inner cities all present different challenges to the system of higher education. Thus, colleges need to understand their mission and (re)define the kind of students they can best serve. Putting the student first requires a flexibility in the academic program to meet different needs created by a range of ages and diversity of the student body.

Putting students first also requires recognition of differences in the stage of maturation of students and in their learning styles. Individuals process the information that they receive in the way they "see" the world. Thus, they develop and use markedly different styles of learning and solving problems. If the student—or faculty member—has developed appropriate competencies, learning becomes a process through which experience is transformed into knowledge (Kolb, 1984; Wilson and Morren, 1990). Because their preferred styles of learning differ, people possess different strengths and weaknesses. The features of the learning style are patterned and measurable, and they reflect much more than differences in background, attitudes, expertise, or viewpoint. They affect the ways students apprehend new experiences, how they grasp issues, and how issues are transformed into knowledge.

Students do change and "mature" in their college careers. The general process of individual and personal development that occurs with maturation is thought be accelerated by exposure to higher education (Korn, 1986; Pascarella and Terenzini, 1991). A useful model of intellectual development and change in learning styles is presented by Perry (1981). Students in colleges of agriculture are little different from students in other colleges in that they enter college with a learning style shaped by dualistic thinking. They divide meaning into two realms: good or bad, right or wrong, *we* versus *they,* what is not success is failure. The professor is the authority. Right answers exist somewhere for each problem, and authorities know them. As they mature, students progress to the point of multiplistic thinking, wherein diversity of opinions, values, and judgment are recognized as valid substitutes for facts and where evidence, logic, systems, and patterns allow for analysis and comparison. Multiplistic learners see that choices and decisions are made in the awareness that knowledge is relative and dependent on contexts. It is hoped that, somewhere in the process, the student develops the capacity for comparing the assumptions and processes of different ways of thinking (Perry, 1981).

Such transitions as described above would seem fundamental to curricular design and teaching strategies. Rules would not be taught or learned as authority, but more as axioms and as the best knowledge available for making decisions. Responsibility and initiative, which were considered by the student as faculty authority, become internalized by the student. A variety of devices exists to bring the needs of today's students into the process of higher education in agriculture. For example, students could be provided multiple approaches to an assignment centering on the same content.

As noted above, the entering student of today is different. In some respects, the typical college student of the future may be more mature in his or her progressive mode of learning. Such students just might be less intent on conciliating authority and begin to learn on their own initiative. Those students in the college of agriculture who come from the more traditional sources (farms, ranches) may be more or less fitted to the cognitive task at hand, depending on whether the institution's program for education in agriculture is more traditional or not. But, increasingly, students come with experiential bases different from the farm, and the workplace may develop other, but yet appropriate, strategies for learning. They will likely ascribe different meanings to learning in different contexts and at different times of their lives, but some students may expect more direction in their courses and will be happier if they receive it (Perry, 1981). Students who wish an atmosphere of greater freedom, who will be an increasingly larger portion of our student body, will be frustrated in courses that are taught more tautly.

Educators cannot coerce students into intellectual and ethical development, even if it were ethical to do so (L. L. Knefelkamp, cited by Perry, 1981). We envision, instead, teaching and curricula designed to invite, encourage, challenge, and support students in their intellectual and ethical development. An undergraduate animal science student at Texas A&M University, after participating in a course on contemporary issues in animal agriculture, observed that students should be taught how to gather facts and do in-depth thinking for rational, philosophically consistent decision making. They should be taught the limits of authoritative knowledge (science). "Colleges should provide a diversity of knowledge, without bias, and teach the students how to come to a conclusion but not force the conclusion itself."

An important part of the relevant university experience is outside the classroom and should be used to build a collaborative learning environment. The educational impact of faculty on students extends beyond the classroom (Pascarella and Terenzini, 1991). The extent of informal contact with faculty is positively linked with a wide range of outcomes. A collaborative learning environment is provided by active departmental extracurricular and noncredit programs. Some of this is provided by both planned and informed social interaction. Such experiences could be fostered by work experience, discovery experience, and encouraging faculty-student contacts by a wider use of guest speakers from the graduate faculty in courses to provide initial contact. Such "rounding out" of the student is a logical complement to systemic change in the program.

The New Paradigm

A new paradigm invites consideration in the hierarchy of education. Critical factors in agriculture and related areas—economic productivity, population growth and distributions, technological development, and the behavior of individuals and institutions in the consumption of products and services—are strongly influenced by the *cascade of knowledge* (Malone, 1994). This cascade consists of "discovery, integration, dissemination, and application of knowledge concerning the nature and interaction of matter, energy, living organisms, information and personal behavior." The slice that a college of agriculture takes out of this cascade is a matter of institutional decision. Niches are involved. Colleges of agriculture generally educate for discovery at the graduate level and for dissemination and application at the undergraduate level. Education for integration remains a niche largely undeveloped in colleges of agriculture and related areas.

The design for an educational system, or portion of a system, rests on at least two sets of premises about resources and needs. One set pertains to the economy; the other set pertains to society and citizenship:

1. Economy: skills, workmanship, incentives, wealth creation and distribution, decision making, management, consumption, and policy.
2. Society: ethics, humanity, well-being, power and political process, community, survival, and development.

The character and purpose, that is, the niche that each educational system targets, will depend upon the character and purpose of the society encountered and the means (economy) with which society's purpose can be achieved.

Other parts of the education a student receives grow out of the special needs of persons to be educated. Career changes are generated by an economy requiring new skills, adaptability, and flexibility. A large part of the labor force, for example, needs to understand computer design and use, robotics engineering and design, biological engineering, and a variety of logic, mathematics, and language skills. Such skills are needed by workers, not just managers. Such education may fall outside the scope of a college of agriculture, although it may be managed or supported by such a college. A reductive agricultural industry may wish to do the training itself in the specific technology that is required by the specific job in industry.

Coulter (1994) made the general observation that there was a time when the missions of land grant colleges of agriculture were so similar that they essentially represented the same institution. Today, the differences exceed the similarities, at least in number. Colleges of agriculture range from the narrow to those with substantial breadth and depth in disciplinary capacity. Breadth and depth represent a great strength for the university, and at the same time, the complexity of it all presents challenges to those who seek to guide and direct these activities. Faculty creativity can be focused across a spectrum of human needs and is often redirected to meet new challenges and research opportunities.

Likely, there will be special educational roles for the colleges. Niche programs should evolve. There is too much left to be done in higher education in agriculture to abandon the special qualities that have been the strength and tradition of the college of agriculture.

Integration of the Entire Land Grant Complex

Departments in the land grant colleges of agriculture are often held together more by state and federal research, extension, and industrial grant funds than by a common bond of educational goals and objectives. Periodically, through the years, some land grant colleges of agriculture temporarily developed some separations of the faculty teaching the undergraduate students and the research-graduate studies faculty. And in some courses, the subject matter appeared to be fixed, belying the often-claimed attribute of the land grant structure that the fresh results of research flow into the courses taught to undergraduates. This situation probably occurred as disciplines became nearly static or so highly reductive that the results of the research had little significance in undergraduate courses. The educational system does a good job of teaching science but now needs to gain an appreciation of how science and technology may drive policy and how policy may drive science and technology.

There will be benefits in integrating research and extension in the teaching program. These represent areas of scholarship on the move, such as genetics and animal and plant modification. Daily new techniques, new angles on old theories, and new genetic products and procedures increase the effect on human lives. Complementarity of education, research, and extension may facilitate addition of such issues as environmental issues, food/health relationships, and food safety issues to the courses in agriculture and natural resources. One may question whether there is any element of education that is static.

The higher education in agriculture, food, and natural resources will continue to serve as a vital part of the research university. The teacher/scholar/learner will be in active pursuit of knowledge and transfer of knowledge in the classroom. New research areas, however, may not be directly transferred to the undergraduate program, in part, because they are reductionist, but more likely because the fresh knowledge is considered to be "advanced" knowledge suitable only for graduate students. But if the work life of the researcher/teacher is varied and challenging, this can enrich what faculty members do and how they think, even though teaching and research become somewhat disconnected. Moreover, the industrial knowledge base enriches itself by integrating new, sometimes unexpected, knowledge derived from science. Rigidly playing to the structure of an individual faculty member's appointment, either as part of the sense of the position description or sometimes as part of the reward system, can have a negative impact on undergraduate and professional education.

A disciplinary research-oriented faculty is "in place" in most land grant colleges of agriculture. Some non-land grant colleges and departments of agriculture may not have the "luxury" of a research faculty, yet their faculties remain largely discipline based.

But in almost any program of higher education, there are outstanding faculty members with seemingly unlimited capabilities. The problem is that these capabilities are not always converted into success in the classroom. And how do we integrate the unique capabilities of scholars within colleges of agriculture into appropriate programs in other colleges of the university?

Somewhere in undergraduate education, agriculture, food, and the natural resources—science and industry and the issues they raise—must be seen as an integrated entity. It is a specific part of society, with its particular historical, economic, and territorial context. Packing learning about animals or plants or foods into disciplines has little to do with how students learn; it has mainly to do with the centrality of disciplines in faculty members' professional lives.

The disciplinary orientation seems inherent in the academic system. The discipline-based structure is not easily overhauled, but it is possible to push against it rather than to defer to it (Alexander, 1993). Paradoxically, lower-division education may provide the starting place.

There are also now professionals—consultants, industrial scientists, managers, CEOs, webmasters, etc.—in the "real world" who would like to be part of the educational programs as integrators of knowledge, and some are remarkably insightful. In the twenty-first century, there will be a wide diversity of opportunities, and it is imperative that higher education look at and educate for an increasing diversity of opportunities. A way to do that may be to call on professionals to enter the teaching programs as part-time faculty. The educational process could change through utilization of extension personnel in the classroom. They can provide case study ideas. They can enhance curricular development. They will not likely take responsibility of student guidance and quality control, but they can enrich the program and amplify systemic change.

Collaborations

Some of the most productive collaborations in agricultural research have resulted from networking, that is, from the informal interrelationships and communicative networks formed by scientists of equal stature finding common interests. The administrative functions in these cases are ones of intellectual and resource support, but generally not of specific administrative direction. Until recently, such networking was relatively rare in academic programs and, if present, seldom extended beyond the bounds of the college. But enlightened faculty members and administrations are beginning to find ways to stimulate networking for academic purposes.

A principle facing colleges of agriculture is that the growth of information is such as to make it difficult, perhaps impossible for a single institution, to expect to meet everybody's wishes. Resources are limited. Thus, again, it seems that the concept of educational niches will become the necessary guiding aspect, but electronic technology is making collaboration far easier.

As institutions identify their niches, new models of educational effort and delivery are emerging that ensure access on the one hand and maximize the effectiveness of expenditures of human capital on the other. Coalitions of institutions both within and beyond geopolitical boundaries are being formed to deliver programs that meet broad demands, particularly in the northeastern region of the United States (Maw, 1994).

Although regionalization, that is, cooperation among institutions across state lines, is one approach, it has not been politically popular in much of the United States. However, the current and likely future resource constraints, coupled with increasing public compulsions for quality improvement and productivity increases plus the availability of digital technology, will compel many institutions and policy makers to periodically revisit cooperative institutional arrangements for academic program delivery. Where numbers of institutions exist within the states, such as in Texas, collaboration can be expected to increase, if but slowly.

As new arrangements are developed involving multiple institutions that might share course work, students, programs, and faculty, the new technologies—particularly in the area of telecommunications—will be increasingly employed. They will be explored as optimal systems for maximum efficiency and effectiveness. They should not be viewed (necessarily) as substitutes for the more traditional models, but rather as complements. The rapidly developing information technologies for communication present opportunities for real-time, interactive distance learning and, as such, can foster the establishment of collaborations that will network providers of information and education with consumers in a variety of new networking systems.

The Image

Faculties in higher education in agriculture will likely expand their vision and scopes. A larger view, perhaps expressed in a new image (name) for the college of agriculture, has engaged biological and functional disciplines and new aspects of human resources to provide what will be perceived to be the full scope of higher education in agriculture. The desire (need) to maintain identification with "agriculture" may hinder the ability to form the needed stronger bonds within both the university and society. But it is clear that whatever colleges of agriculture do, they must begin to appeal to a larger population of students and maintain credibility across the campus and across the boundaries of city, county, and state.

The land grant concepts of accessibility and service should encompass the whole university. The colleges of agriculture may have a special responsibility to this end. The relevant field that is most interactive with society is food. In the elements of food and nutrition, problem-solving skills and an understanding of basic scientific concepts can be combined. But, taught as an interaction of agriculture with society, food safety and nutrition are topics of wide interest among students throughout the university and among the public as well.

The student recruitment problems of the twenty-first century will not likely be the result of the negative image of "agriculture" that plagued colleges of agriculture a decade ago, although some retain images of farmers, ranchers, and foresters as despoilers of our resources. To the majority of teenagers, agriculture has no image at all. That is an advantage: lifted of the burden of the equation of agriculture with farming, colleges of agriculture can start the twenty-first century with nearly a clean slate. But that will demand a reincarnation of the subject as one intellectually interesting and worthy of attention as a pathway to a career. The students of tomorrow will be drawn to the college because of the educational opportunities provided, not because the curricula are housed in colleges of agriculture and natural resources. Colleges of agriculture will recruit students more likely as a result of their interest in living things, the environment, nutrition and health, the rural landscape, and service to humanity than their interest in agriculture and its industries. Likely there will be other interests even more telling in the twenty-first century.

Conclusion

Multiple issues require consideration in developing a vision of higher education in agriculture. They involve evaluation of programs in terms of planning for excellence, determination of the courses and learning experiences, and structuring the learning experience to meet the criterion of excellence. The relevant issues include questions of how faculties view students as they arrive on campus: as potential technicians, potential job-holders, potential entrepreneurs, or potential leaders. Whatever view is held can be fulfilled by appropriate education.

The sense of excellence in education incorporates the concepts of good, motivated teachers, well-equipped classrooms, and a supportive administration. It also includes individual faculty members engaging in the "scholarship of education," which includes development of innovative learning situations and self-improvement in teaching by exploring innovative teaching methods. The student will engage in his or her own learning. The faculty member will join the student in learning. It seems axiomatic that the colleges of agriculture should tailor their programs to meet the needs of the students that they admit, not the convenience of the faculty and staff.

Putting the student first is as much a pervasive principle of thought as a practice. Colleges of agriculture will have to focus education to serve students recruited out of the diversity of populations in the United States. Putting the student first includes recognition that education should develop (mature) the learning style of the student, not simply respond to the differences of existing learning styles, although such response may be the vital starting point.

The vision for higher education in agriculture will no longer be based on a common purpose for each institution. Even in colleges of agriculture, or especially in agricultural colleges, the concept that each college has its niche in higher education should be a guiding principle.

Many of the units of higher education in agriculture are in land grant or research universities. The vitality of the program in higher education in agriculture is enhanced by the integration of the research and extension personnel into the undergraduate program. This conscious development enriches both the knowledge base of the educational program and the academic environment, by adding diversity of thought and providing the view that agriculture, food, and natural resources and science and industry form an integrated entity.

The future college of agriculture will not be a unit that stands alone. As agricultural research has developed its networks, so should higher education in agriculture. Coalitions of institutions will likely be formed. The electronic campus will be in the new framework. The college of agriculture will likely not only be a professional college, but a liberal college as well, increasingly relating to and serving students across the campus. The college of agriculture can be envisioned to be a center of scholarship, capturing the interest of students in living things, the environment, nutrition and health, the landscape, and service to humanity. It is a center of scholarship that translates its vision into the educational process through curricular and programmatic design and development.

4

Forms of Higher Education in Agriculture

Cameron Faustman, John Riesen, Dwayne Suter, and Donald Vietor

Key Concepts

1. New models of education may be among the most useful frameworks for revolutionizing higher education in agriculture. Other models have new applications. The educational environment is changing within universities, creating both challenges and opportunities.
2. New forms of education include changes that are delivery based (modular courses, overarching themes in course content, blending theory and skills), changes that are reactions to changes in audience (for example, applications of science, ethics, and life-long education), and emerging delivery technologies (PowerPoint, the CD-ROM and digital video disc formats, and the Internet).
3. Virtual expansion of the campus through electronic technology offers significant opportunity as well as challenge.
4. Education of the freshman student offers an important point to initiate a revolutionary education in agriculture and natural resources.
5. Other forms of education that were explored in the first edition of this book (Kunkel et al., 1996) could not be validated as being feasible in the system of higher education in agriculture in the foreseeable future. They are not included in this edition, although this may be a premature judgment.

Introduction

As the programs of colleges of agriculture are in flux (Doering, 1992; Handelsman, 1992), innovative curricula and new forms of higher education in agriculture are being proposed and tested. The fundamental requirement for each change is that there be a definition of the intellectual purposes of the program. Any redefinition of purposes requires rethinking the curriculum.

Elements of higher education in agriculture in the future will likely incorporate different definable bodies of knowledge and require different forms of education. Themes or areas of domain-specific knowledge that have become appropriate for higher education in agriculture are sustainability of natural and human resources, sustainability of food production, international agriculture, sustainable biodiversity or conservation biology, environmental agriculture, process agriculture, health and food safety, industrial agriculture, biotechnology in agriculture, regulatory liaison and technology policy, and regulatory science. Each element may find a different level of implementation. Industrial agriculture, for example, may be the course of study such as the current poultry science curriculum seems to be, an industrial option of animal science, an agribusiness option in agricultural economics, or an industrial (agricultural) waste emphasis in soils.

New models in agricultural education may stem from redefinition of the professions of agriculture. Expectations of graduates are changing. Some clientele of higher education see needs for developed people skills: abilities to negotiate, communicate, and become change agents and leaders. The ability to function as an effective individual regardless of the subject is seen to be critical. Some clientele see needs for graduates with broad interests and the ability to integrate. Parallels with new problems-based educational thrusts in medical education are suggested for higher education in agriculture. Inclusion of experienced practitioners in the process of higher education should receive consideration. Commitment to the system is urged. Alternative forms—new approaches—in higher education in agriculture may be the most useful framework for systemic changes.

Concerns for Newer Forms

Some new turns in higher education will require upgrades in technology, while others modified versions of what has already been successful. In any case, each can be an addition to the tools that skilled teachers have at their disposal. There must be some courses that provide significant exposure to problem solving, critical thinking, and experiential (hands on) laboratories. These will be more costly than their traditional counterparts in terms of financial resources and preparation time. Thus, other courses may have to have larger enrollments, be taught by graduate assistants, or use other cost-cutting methods; the balance must be determined with the student in mind.

Alternative forms of education can address subject-based, delivery-based, and audience-based concerns. The challenges presented by these authors are applicable to agricultural higher education and are addressed here in the broad context of systemic change.

Subject-based Concerns

Conventional wisdom suggests that subject-based materials will usually include assimilation of current and emerging agricultural concerns within existing curricula. These generally fall within the concept of domain-specific knowledge or themes and include such things as agroecology, sustainability, and internationalization, but it is unlikely that all of these areas can be taught as stand-alone courses, especially in smaller schools. A logical approach to implementation of systemic change is to incorporate the modern concerns about agriculture into the fabric of existing courses offered across departments within colleges of agriculture.

The origin of tenure was to protect academic freedom and allow instruction of material regardless of political or other pressures. However, there should be a recognition that undergraduate education must evolve to meet the needs of a changing world. Faculty will need to "buy into" the concept of new components of higher education in agriculture. In a sense, faculty members may themselves require education about broad-based agricultural concerns and may need to be shown a successful example in practice.

A substantial challenge to undergraduates with regard to subject-based concerns is related to the expanding base of knowledge. Emerging technologies do not negate the need to learn fundamental principles, many of which have not changed for decades, but recent discoveries provide different material that must be added for an understanding of where the field is today. What that material is is not clear. This is not a problem specific to colleges of agriculture but applies to all science-based curricula. However, it seems unlikely that expanding the knowledge base will require more time for undergraduates to obtain their undergraduate education such that the four-year B.S. degree would become a thing of the past. The four-year undergraduate curriculum is here to stay, for political reasons, even though many students today use five years to complete it. Restructuring course content to present mainly the elements of knowledge essential to the desired learning outcomes, instead of "covering" the entire subject content, will likely be one of the most important new forms of education that can be adopted.

The principal deficiency of higher education as perceived by students throughout both secondary and higher education, including students in agriculture, is the lack of connection among the several courses. Courses that are part of the agricultural curriculum typically lack coherence. Knowledge has often been taught for its own sake, rather than the knowledge that ripens into wisdom or that serves larger ends. Thus, the basic needs of colleges for teaching competence and also motivating students to careers in agricultural sciences will not be well served by the traditional programs of colleges of agriculture. Something more is needed.

Delivery-based Concerns

Delivery-based concerns are pertinent to all curricula within the university but have special significance to colleges of agriculture. This is because the concepts of agriculture and the application of agricultural principles differ substantially in institutions in different geographic areas across the United States. Within a college, delivery-based systems that incorporate multidisciplinary approaches will likely provide a broader view of the subject matter at hand. Such models, while not generally new in the literature on education, have important new considerations in higher education in agriculture and natural resources.

Course Modules

Modular courses generally cover controversial or special subjects or themes and are taught by a variety of faculty. Each faculty member discusses his or her own professional view of the problem. Modules may be obtained from other faculty from other universities or may be downloaded from the Internet and incorporated in the course by the course leader. If developed properly, the course should provide the student with an understanding of the complexity of an issue and a basis from which to make informed judgments. It is an approach that requires coordination. One faculty member may assume responsibility for a given modular course and ensure that the various individual lectures and assignments contribute to a final product worth the effort.

The modular course then is more than a collection of lectures. An appropriate analogy is that of the textbook in which each chapter is written by a different author. Modular courses need a good "editor." The support of departmental and college administrators is very important if this approach is to succeed, as there is a belief held by significant numbers of faculty that time invested in this kind of team approach is not of as high a status, consequently rewarding the traditional single instructor courses. The challenges of this type of course escalate as departmental and college lines are crossed. However, the rewards of an integrative collegewide course examining, for example, agricultural production on a systems level with faculty from areas such as animal science, marketing, ecology, philosophy, patents and intellectual property, soil science, food science, biochemistry, water resources, and geology could well be worth the effort in terms of the intellectual growth of students.

Capstone Courses

Increasingly, faculty members are viewing capstone courses as excellent means to promote learning that interfaces and integrates diverse knowledge garnered from fundamental and functional courses, thereby contributing significantly to the quality of education. A case-study approach similar to that used in law and business schools is probably one of the best methods for structuring a capstone course. The overall goal must be to challenge students to apply what they have learned to solving potential conflicts that typify those they will encounter in their careers. Significant challenges are presented to both instructor and student. As noted by Carter et al. (1990), capstone courses should not "rehash" introductory-level material but should be taught in an integrative manner. Faculty members must have the experience and breadth of understanding necessary to teach these courses. Senior faculty will likely be in the best position to initiate these in the context of what has been historically considered and what will likely be of future concern. The product will be measured in terms of nontechnical skills (communication, problem solving) rather than mastery of technical, discipline-specific facts. Capstone courses can reinforce a knowledge of available resources and the ability to use them to obtain data pertinent to the problem at hand.

Decision cases have been introduced as an incremental new form of education in discipline-oriented capstone courses (Simmons et al., 1992). Decision cases have been developed using the Harvard Business School model. Decision cases are distinguished from the other case studies by an insistence that "a decision . . . must be made . . . even though more data would be desirable and no perfect solution seems to exist." The advocates of decision cases argue that case materials make it possible for students to think purposefully. They conclude that the more students are prepared to cope with the stuff of decision making in the real world, the better professionals they will be.

A note of caution is in order. The capstone course that focuses on a narrow scope of the subject could work against the flexibility desired in the graduate. An example of such a narrow focus is the "capstone" animal science course that may integrate soils, forages, nutrition, and reproduction, but with the focus on only one segment of animal agriculture such as the cow-calf operation.

Themes

An approach worth capturing is that central themes can and should cut across courses. The greatest utility of a theme may be in teaching core principles rather than detailed knowledge of the mechanisms of agriculture. One such theme is embodied in the question, What would we like to have a student in agriculture be prepared to do about social and ethical questions? How can students develop into responsible policy makers and leaders for agricultural entities? Doering (1992) argues that higher education cannot avoid the responsibility of trying to answer the question by using the device of throwing our students into international or other cross-cultural experiences to sensitize them to "real" issues. Students must first be able to identify and deal with the context of their own society.

Many faculty members in agriculture believe that it is important that students know that science is a process of discovery and that the education of agriculturalists should be conducive to inquiry. That too can be an overarching theme in courses (Williams and Young, 1992). But the relevance of research to daily life is often not perceived by the student. Students see science courses as stumbling blocks on their way to a degree. Science courses are generally not designed with creativity in mind. Scientific papers are dull reading for students. Part of the answer lies in balancing the inquiring process with the knowledge content in our courses. A focus on the inquiry process can be achieved in several ways, but the structuring of a course, say, a laboratory course, that progresses from performing the traditional by-the-book experiments and exercises to asking and answering questions and to small groups designing and executing their own experiments may be one approach (Eisen et al., 1992).

Blending

Theory in higher education in agriculture suggests that the undergraduate should blend knowledge and skills. Drucker (1994) suggests that the educated person required in the work world of the future is one who has developed both specialized skills and theoretical knowledge. He suggests that a person educated for the work world of the future should not be educated as a generalist, but as one who can adapt to specialized techniques and also be able to acquire new specialties rapidly in order to move from one job to another.

Skills such as the ability to maintain an extensive vocabulary, use computers, make appropriate use of syntax in writing, feed and water a plant or an animal, and judge a situation critically blend with the ability to learn and use knowledge. Curricula of tomorrow must include both the studies of processes and such qualities as learning to write, speak, and listen, the ability to work in teams, and leadership. This blend of knowledge and skills ranks high among employers as desirable characteristics of college graduates.

Hands-on learning also provides understanding and intuition that takes principles into reality. In blending and developing themes, it will be both the faculty member's and the student's responsibilities to take ownership for the course. Both players must agree to search for a common thread that connects abstract knowledge to application. The faculty members may be able to work their courses as in teams. If the theme is "environment," the faculty in consultation with each other can find ways to incorporate the topic into their courses. This technique includes a timing device that may allow classes to merge occasionally for a special presentation or experience.

The curricula in colleges of agriculture should provide for fundamental learning, instrumental knowledge, and knowledge of the processes of science, food, natural resources, and agriculture in their broadest definition. It is in thematic and skill blending that students can make connection to reality and among courses.

Experiential Learning

The concept of experiential learning is not new but should be given new attention. Hands-on experiences are favorable learning methods, but the concept of hands-on educational experience must change in the light of what is considered important in the work world. Thus, experience in an agricultural farming practice would be replaced by quality control and computer programming for management of an agricultural operation in industry.

Students often respond in a positive manner to experiential learning opportunities that are well organized. These opportunities usually occur at nontraditional (nonclassroom) locations that provide high visual impact. The major constraints are effective class size (small is usually better) and costs per student (certainly higher than lecture only). Many faculty members have abandoned design of these because of the substantial preparation time required, and/or because new faculty members in many cases have not themselves been exposed to experiential learning activities.

As noted in the earlier chapters, experiential learning opportunities can be provided via internships and cooperative learning opportunities.

Emerging Delivery Technologies

With the development of various disciplines, communication beyond the local academic community developed through letters, manuscripts, and publications. The use of documents became integral to organized and formal education. The granting of academic credit for the completion of a specified program of study was initially limited to work completed in residence. The offering of correspondence courses is one of the oldest and most extensive models designed to provide access to campus-based academic programs from remote sites.

The use of documents remains important in higher education, but the delivery can be by electronic means, essentially providing a new form. Electronic and communication technologies are making rapid advances. These advances have changed the form of education that is possible from both technical standpoints and economic feasibility. Delivery systems that allow direct instructor-student interaction (e.g., PowerPoint technology) and personal computers that can be responsive to students' individual learning styles now provide unparalleled opportunities for learning. Without student-instructor interaction, however, the experience can be frustrating to the student.

The introduction and utilization of *electronic* technologies expand the "field of play" where learning takes place. In fact, in cases where multiple distance education sites are utilized, a wide variety may exist in both the field of play and the "players." Strategies need to be employed to accommodate the variation in learning styles existing at the sites. Experiences in distance education thus far have pointed out the need to have a local site facilitator at each location to answer questions, lead the discussion, and focus on the learning process. In several cases where no local facilitator was provided, the distance education program was unsuccessful and was soon canceled. It will be important that all players directing or managing the learning process understand, and are experienced in, the design and delivery of educational programs that recognize the wide variation in learning styles, and that they plan accordingly. Direct human contact still seems necessary. If not necessary, it is certainly preferred by the student.

One adaptable technology is the CD-ROM format. Digital versatile disc (DVD) developments offer further application. The CD-ROM is a computer-coordinated module that is capable of using high-quality images, sound, video, and text. CD-ROMs have been used effectively to present case studies, allowing students to identify the problem, request specific information, analyze the information, and recommend a course of action. Used appropriately, computer-aided instruction has increased learning efficiency, retention, and depth of understanding. Feedback is generally offered based on student responses for each step. Students can proceed at their own pace and at a time that fits their schedule. Modules can illustrate complex concepts using animation or simulations. Students, in turn, can experiment with the principles, say, of genetics. Genetic experiments that would take many years in real life are simulated with the computer in minutes. Students can test hypotheses they create and "discover" basic relationships and concepts. Still other modules are able to provide illustration of complex relationships and structures with three-dimensional animation or video footage. The student can slow down rapid events or speed up slow ones to better visualize processes.

The Internet and World Wide Web are noteworthy developing technologies in electronic communication. Local area networks (LANs) and the Internet can be used as forms of distance learning, but each also offers great possibilities of accessing information in an independent manner. In significant numbers of residential universities, the Internet may be more useful in on-campus instruction than as a mechanism of distance learning. As students access the Internet and other networks, class discussion groups have become feasible. These systems have been used successfully by faculty among themselves, but at the level of current development all members of a class can receive the questions, answers, and/or comments from each other. Questions might be trivial, such as when the next assignment is due, or they may deal with philosophical issues related to the course or with clarification of a concept. Students read their electronic mail (E-mail) at times and places convenient for them. The instructor can monitor the discussions and contribute when appropriate. Using the Internet, this same type of activity could be carried out at a national or international level. The Internet allows one to make a telephone call, watch a video, listen to an audio broadcast, broadcast oneself, shop, learn, and communicate. E-mail has turned the network into a new communications link. This is clearly a form of education. The Internet will likely provide the largest amount of new knowledge available to the student in the twenty-first century.

Challenges

All of these forms of delivery present challenges. As these electronic technologies emerge, they may challenge the established formulas for course credit. In terms of student learning, is it possible to have one formal discussion hour and abundant special "out-of-class" projects plus additional E-mail discussion equivalent to a standard three-hour lecture course? Just what does earning one credit with a grade of B imply?

There are many three-credit courses added to a student's plan of study because they are deemed to be prerequisites to a course of major importance. Closer examination will often indicate that only a relatively small portion of a prerequisite course is necessary to the understanding of the subsequent course material. One can envisage a series of small, relatively independent modules that in total would embrace the content of several related courses without the redundancy now common. Each module might have strict prerequisite modules. Instructors would not be forced to choose between letting students take their course without some of the prerequisite preparation and forcing them to take a full three-credit course for a relatively few concepts. By choosing the appropriate set of modules, students could build a customized course that meets their specific needs while allowing for efficiency of larger numbers and elimination of duplication in presenting common concepts. Such freedom, however, should be handled with caution; it may encourage specialization or reductionism or diminished rigor. The lack of quality control of the material on the Internet may be a challenge to students and faculty. But the Internet may be the principal source of information in the twenty-first century. Education then may be a matter of integration.

Elimination of traditional boundaries and the development of new networks, alliances, and partnerships are occurring. The speed at which further changes occur will reflect further innovation.

Audience-based Concerns

Colleges of agriculture must actively seek to reach the broader public. There is a need to become involved with continuing education/life-long education of individuals in all fields. The mission should be defined, and it should be recognized that it is distinct from that of colleagues in schools of arts and sciences, engineering, and others. The application of fundamental science principles provides the unique opportunity to make people aware of how science affects their everyday lives. Colleges of agriculture need to educate those who educate others. Although this has generally occurred for teachers of agricultural science and technology in secondary schools, there have been few attempts to teach teachers of earth science, biology, and chemistry about how agriculture, food, and natural resources embody the basic concepts embraced in their discipline.

Colleges of agriculture need to communicate their scopes and their effective forms of education to the general public. The community should be more evangelistic about informing the public both about food and natural resources and about the importance of agricultural colleges to society.

The Electronic Campus

New technologies will create a different kind of campus. The electronic campus is one of partnership and collaboration (Pacey, 1993), which will challenge higher education in agriculture and related areas.

The electronic campus is not limited by brick and mortar on a physical campus but is more a network of educational providers and learners. Telecommunication technologies provide the linkages between educational providers and learners in both real time and asynchronous modes. Interactivity, therefore, can occur in real time or at the convenience of both. This interactive capability and rate of development of effective telecommunication technologies are enabling the rapid expansion of distance education programs.

Effective use of distance education technologies will provide access to larger markets. Geographic, economic, political, cultural, and other boundary conditions will have less limiting effects than currently exist. That will create some problems. State agencies, governments, institutions, and other gatekeepers are already struggling with the issue of citizens having access to delivery of a wide variety of distance education programs. Issues such as out-of-state tuition, admission criteria, and validation of academic credits awarded require creative thinking and resolution.

But, clearly, linkages across state and national boundaries will lead to global markets for United States institutions of higher education. Access by some institutions to one or more of the global markets already exists. The Internet will play a major role in institutional and individual access to global markets. Entrepreneurs outside academia will develop needed educational modules for specific market niches. The Internet will play a major role in providing access by the learner to the modules, completion of assignments, validation of achievement level, payment of fees, and issuance of an electronic certificate of performance.

The Internet is the most rapidly expanding worldwide computer network. It is literally a "network of networks." On a given day, several million users in more than fifty countries are connected through the Internet. People are given the opportunity to connect between offices, between faculty and students, and between and among people throughout the world. It is a ubiquitous open networking. It will become simply a way of life for institutions of higher education, public entities, firms, and individuals.

An extensive communication system capable of transmitting voice, video, and data is in place now, long before many higher education institutions will likely be ready to make effective use of such capabilities. Digital technology will be the primary method of encoding, storage, and utilization of information in the future. Combining video, voice, and data in high quality multimedia presentations designed to meet different learning styles is now possible but is not understood by most educators. Extensive professional development programs are needed to enable educators and users of the educational systems to effectively take advantage of the current and future opportunities.

Institutions will likely be more limited by their vision than by either technology or finances. What is greatly needed by the users of the future distance education models is a shared vision to which colleges can commit time, resources, and expertise.

Effective distance education requires quality instructional design and planning. The planning process must consider the learners' learning styles, access to technology, variety of educational backgrounds, motivation/need to participate, and several other factors. A paradigm shift from a focus on the education provided to an emphasis on the learner is needed. Selection of technologies and delivery process should be made with a focus on the learning process. This shift is especially critical for continuing education and certificate programs. Employees frequently comment that they wish to buy only certain course modules and do not wish to waste their time or money on the others.

Learning in the "knowledge age" will be facilitated by linking learners in a wide variety of physical places through an atmosphere of virtual learning resources and processes. Many of the resources will exist only in an electronic form. The learning process will be enhanced by interactive linking of learners, experts, and mentors in a variety of patterns of communication.

The potential for a combined impact of new communication technologies with the rate at which new knowledge is produced will be a significant factor in fulfilling the mission of research institutions of higher education. More rapid and effective access to new information and knowledge will be important to both the researcher and the teacher. Collaboration among faculties on a global basis in both teaching and research initiatives will be facilitated through the use of the Internet and, in some cases, satellite delivery of certain segments.

The rapid development of new knowledge and communication technologies has major implications for the publishing industry. Many electronic, peer-reviewed journals already exist, and many more will be developed in the future. Libraries will increasingly focus on the use of electronic storage of holdings because of the cost of purchasing hard copies, decrease in storage facilities required, and ease of access by users. The review process of journal editors is now conducted in many disciplines electronically, with the potential for lowering publication costs and decreasing the time from submission of the article to its publication. Also, in many cases, the program and proceedings of professional meetings can be viewed on-line or by viewing through the use of a CD.

New technologies currently being developed and others yet to be conceived will have a major impact on the electronic campus and those being served by the faculty and staff. Paradigm shifts in how knowledge is utilized, by whom, where, when and why will be a dynamic process in the future. In many cases, the home and the workplace may be the same location for much of the week. Therefore, new technological possibilities will continue to cause shifts "in where we work and where we learn, creating an important new role for the home" (Poley, 1999).

The latent effect of electronic commerce (e-commerce) on the electronic campus in the future, the purchasing of products and services on-line through the use of the Internet, is unknown but is assumed to be of great importance. Currently, marketing of products and/or services utilizing an Internet Web site is growing at a very fast rate. In the future purchasing of educational programs through the use of the Internet will be accepted routinely by many as simply another product.

Education of the Freshman Student

As part of the systemic change, institutions need to capitalize upon the strengths and skills freshman students bring to the college today. In theory, education is built on existing knowledge and experience of the student. The beginning course in an agricultural science was historically based on the incoming student's experience. Other courses in the freshman curriculum were built on knowledge obtained in precollege schooling, such as it might be. The students of two or more decades ago grew up on the farm or ranch. For example, they knew animals and had an affection for them. They knew how animals thrived, reproduced, and were prepared for market, and the beginning course was fashioned accordingly. But they did not know industrial agriculture or the risks of the future. The beginning course in animal agriculture of the future will require another vision.

The beginning levels of college education are a period of new discovery. The challenge is to recognize that it is discovery of what might be, of what really interests the student, of what is available academically, and of what a student's drive and capability can handle. New theory might suggest that the first course in agriculture in higher education be a course in agriculture, not agricultural science, generalizing from the experiences that *most* incoming students will likely have with agriculture regardless of their demographic origin, that is, food, natural areas, and companion animals. The springboard could be a broad inquiry of how food, plants, and animals are provided to different societies and how our society has industrialized agriculture unlike the scientific agriculture of much of the rest of the developed world.

The beginning animal science, agricultural economics, or agronomy course could become more of an issues-specific course than the current one, which is based on factual information alone. If one accepts the principle of diversity, that there are different ways to accomplish the same end, it may not be necessary for all colleges to structure the beginning course in exactly the same way. The focus would be animal agriculture, for example, not disciplinary animal science; food and agriculture, not agronomy; and agribusiness and trade, not agricultural economics. The new demands for management of natural resources and the growing attention to environmental and health problems could be introduced in such courses.

Again, there are faculty members who find disciplinary boundaries constraining and who want to find colleagues with whom they can discuss broad intellectual topics and contemporary concerns with the framework of ideas. They can teach such beginning courses.

Conclusion

Different forms of higher education in agriculture emerge as new dimensions of education are envisioned. The forms of education may relate to the subject matter. For example, different themes (definable bodies of knowledge) are evolving and are relevant to food, agriculture, and natural resources. These themes can include such elements as sustainability of resources, industrial agriculture, health and food safety, biotechnology, and in the international sense, the sustainability of society. Such domains of knowledge may be included as stand-alone courses or integrated within existing courses but require a change in perception of the content of the curriculum and the philosophy of the education. Similarly, the incessant growth of the knowledge base places pressure on the content of the curriculum, and restructuring will be required to maintain the four-year undergraduate program.

Different delivery systems offer the opportunity for incorporating the needed multidisciplinary approaches. Development of modular, capstone courses, and overarching themes, new attention to experiential courses, and the use of emerging delivery technologies will enable approaches to education. A diverse audience presents opportunities for higher education in agriculture to develop systems to reach out with courses of new construction to educate those who are not enrolled as well as those who are enrolled in degree programs in the colleges of agriculture.

Extending the use of such forms as these can provide a variety of prospects or alternatives in systemic change in higher education relative to agriculture, natural resources, and related areas. A scenario for each such prospect can be developed. The prospect of, say, developing the intellectual team of faculty and students as the working unit in higher education provides opportunity for further exploration.

Colleges of agriculture and natural resources are likely to remain residential colleges, not displaced by virtual distant education, because excellent students will seek direct, face-to-face contact with the faculty and the institution. Colleges or departments will be displaced by quality distance education only if their programs become pedestrian or remain out of date. But the various forms of education will transform the campus and its outreach, and they will have their major impact on the education at universities that strive for excellence.

5

Making Change Decisions

A. Gene Nelson and Suman Singha

When all is said and done, will more be said than done?

Key Concepts

1. The most difficult step in systemic change is implementation.
2. Systemic change is institutional change, and faculty members tend to regard institutional change as of less concern to them than matters that are incentives to personal improvement. Incentives to faculty to participate in the change process seem necessary. Moreover, many of those who are interested in systemic change will have personal opinions of what should be done, opinions that often do not have the benefit of debate.
3. Administrators have strategic importance as they can create the environment, support the planning process, and muster the incentive for supporting change. But, ideally, change in curricula ultimately should be a bottom-up process requiring faculty involvement and ownership.
4. Essential to implementing change is understanding what is achieved in the process. The understanding should encompass the framework, the desired principles of future academic programs, and the concepts that direct what ought to be changed.

Introduction

The foregoing chapters have developed a number of important propositions:

1. Agriculture is central to several matters of great concern—food, health, and environment.
2. Higher education in agriculture is damaged by the too common image of an industrial giant concerned only with production of food and fiber, with little regard for environment and natural resources. Too few realize that the context and circumstances of agriculture, food,

and agribusiness have changed drastically over the past three decades and can be expected to change even more in the decades ahead.

3. Collegiate curricula in agriculture are overburdened, with too many courses that have too much complexity, too much technical content (facts and figures), and too much redundancy—but not enough connectedness to one another and not enough apparent relevance to the "real world." But underlying principles of science remain important components.
4. Faculties in colleges of agriculture are research oriented and increasingly are reductionistic and narrowly focused in their scholarly perspectives, to the detriment of innovative, integrative educational programs.
5. Colleges of agriculture are positioned to lead an intellectual and ethical revolution that will help improve the human condition on a grand and global scale.
6. Systemic change in agricultural higher education is essential for its continued contribution to the sustainability of society.
7. Effecting change in the content and the teaching of curricula in colleges of agriculture is the challenging proposition.

There are two sources of change. The first is internal, in which college administrators and faculty seek to initiate change as a way to control the future of their programs. The second source of change is external, imposed from the outside by higher authorities within the institution or by the institution's external environment. Examples particularly important for land grant institutions include changing amounts and sources of funding, changing student characteristics and numbers, changing expectations of their future employers, changing and growing opportunities for careers of graduates, and the changing clientele of agricultural research and extension programs.

The market conditions under which educational institutions operate have been important factors influencing change (Hefferlin, 1972). Departments and colleges must attract resources to thrive. Departments and colleges now compete for resources, both inside and outside the institution. Departments and colleges compete for students, new faculty members, operating funds, and research contracts. In this marketplace, academic units watch what analogous units are doing at other institutions and try to emulate their success.

The source of change influences how faculty (and other people) respond and accommodate change. Budget crises in publicly funded higher education institutions in the past have imposed change at an intensity that few find comfortable. According to Rush (1992), higher education's most common reactions to budgetary shortfalls have included the following imposed changes: hiring and spending freezes, across-the-board cuts (usually by some percentage amount), and reductions in force (RIFs), or layoffs. The problem with these changes is that they are often temporary solutions, carried out without a long-range plan or vision. Typically, little or no consideration is given to the impacts on the quality, priority, or viability of programs. As a result, these changes are often damaging to the institution and frequently harmful to faculty morale.

Change imposed from the outside has obvious negative implications because people do not like to be told to do something, even if they agree with it (Newton and Tarrant, 1992). When change becomes reactive with a crisis-management mentality, the capacity for self-motivated or self-imposed change is reduced. Individual faculty members lose their vision of the potential of their departments and programs. Pessimism becomes a prevailing attitude.

Thus, to stay in control of the change process, faculty and their administrators must be aware of the changing environment, anticipate the future, and develop plans to change accordingly. The change process should be internalized. Excessive externally imposed change has negative implications and discourages faculty involvement in programs and curricula.

A proactive, self-imposed change process requires faculties to rethink their current situation and the future potential of their departments. Faculties must operate in the context of a long-range plan if they are to make significant changes in their courses and teaching methods. Rush (1992) writes, "The challenge is not merely to find a way to survive with fewer funds in the future, but how to thrive with reduced budgets." He argues that the emphasis in this planning process should be on improving quality rather than on increasing productivity. It is the administrators—deans and department heads—of colleges of agriculture who must take ultimate responsibility for changing academic programs in agriculture. They must decide if they, with the faculty, wish to address the need to reorient the teaching programs in agriculture. If the answer is yes, these administrators may wish to make themselves into change agents, that is, initiators of change who affect the work environment of their faculty. This suggests that department heads' performances might well be evaluated based on their ability to enhance the total academic performance of the department by helping their faculty members do their jobs (Meyer, 1993). Change is the means by which colleges of agriculture and their departments can take control of their futures.

Managing change involves two challenges: (1) deciding what changes should be made and (2) deciding how to get maximum acceptance from those who are involved. This chapter discusses the processes involved in answering the first question. The second question is discussed in part 2 of this book, which is on implementing change. To facilitate change, the appropriate combination of incentives must be developed. Funding sources must be developed, the need for change communicated, and intellectual excitement stimulated.

A Model for Managing Change

Kirkpatrick (1985) outlines a seven-step model for change. These seven steps represent a systematic approach for deciding on the needed direction of change and then managing the change process:

1. Determining the need or desire for change. In strategic planning, the need for change is identified through the environmental scan, an assessment of strengths and weaknesses, and exploration of opportunities as well as the expectations of society.
2. Preparing tentative plan(s). Plans to implement change should be tentative, meaning that they are subject to change. Introducing the plan as tentative encourages faculty to suggest alternate solutions and modifications and to provide their reactions before the plan is finalized. Visioning should enter the process here.
3. Analyzing probable reactions. Some faculty will resent and possibly resist the change. Others will remain neutral. At this step it is important that administrators understand the individuals who will be involved in the change and be able to empathize with them. To accurately assess the degree of resistance or acceptance to change, it is necessary to consider each person individually. The better the administrator knows the individuals who are affected by the change, the more accurate will be the analysis of their reactions to it. By understanding why some resist change while others accept it, the reactions to a contemplated change can be anticipated.
4. Making a final decision. This may involve either a unilateral decision by the administrator or a group decision. The unilateral approach is quicker and emphasizes the authority and status of the administrator. This approach can be effective if the administrator is trusted and respected. On the other hand, the acceptance of change is enhanced when the people who will be affected can participate in the decision-making process. Whatever the approach, the key is to arrive at the best possible decision with a high level of acceptance by those involved. Participation requires the administrator to involve those who will be concerned about, and affected by, the change. According to Kirkpatrick, participation contributes to the quality of the change and increases the acceptance of those who must change. This is what "managing" change is all about. Managing change involves both the decision itself and its implementation. A good decision can fail because of lack of acceptance, resulting in resistance and even sabotage. Seek the reactions of those involved before the decision to change is final. Listen to them and consider their opinions as well as the facts in making

the decision. If suggestions are not used, explain why not, and give people credit for those that are used.

5. Establishing a time table. The schedule for carrying out the change may be as important as the change itself. If there is little resistance and everyone is eager for the change, then the change should be implemented with as much speed as is practical. When resistance is strong, however, the change should be carried out slowly, possibly with a pilot program to demonstrate effectiveness.
6. Communicating the change. This continuing process actually begins with step 1 and needs to be two-way, both presenting the plan for change and listening to reactions and suggestions. After the formal decision has been made and the timetable established, however, a coordinated communication effort is needed to facilitate implementation. Communication means creating understanding, not just telling people. The information must be communicated to those who want to know, as well as those who need to know. Timing is often crucial; people should be notified as far in advance as possible, but timing of the communication should also coincide with the need to know. One approach is progressive disclosure with general information provided early and more detailed information provided closer to implementation. Usually, both oral and written communication may be necessary to achieve understanding and gain acceptance. One advantage of oral communication is the opportunity for feedback. Let people know as far in advance as practical. Explain why the change is being made, what is involved, and how it will be accomplished. Seek feedback to be sure that people understand.
7. Implementing the change. This is the action step in which the final decision is implemented according to the established timetable. Implementation also involves evaluation to assure that the change is proceeding as planned and, if not, to allow for appropriate adjustments.

Critical elements in this seven-step process include empathy, communication, and participation.

Creating an Environment Conducive to Change

The role of administrators is to foster an environment that encourages change and innovation. The essence of this role is to work with the faculty to provide a vision, mission, and plan to guide the change process.

Synergy refers to the achievement of cooperative action with minimal effort. Synergism, according to Siebert (1982), produces "dynamic, although often unexpected results with little effort, creating a whole that is greater than the sum of individuals." Thus, although competition provides impetus for innovation and change, it is complementarity within the organization that facilitates this change process. These complementary relationships, in the language of organizational behavior, are described as synergistic. In a synergistic organization, less energy is required to accomplish tasks.

Achieving a high level of synergy involves both administrative encouragement and individual faculty initiative. The administrative structure must provide incentives for cooperation and set standards for evaluating performance. Individual initiative begins with faculty members defining their own self-images (i.e., a good understanding of their strengths and how they can contribute to the department's mission) and then developing interpersonal relationships of trust and respect with other faculty members.

Planning for Change

Planning creates the environment for inducing change. The process of planning in academia should be continuous and developed such that departments and faculty are informed and committed to carrying out the plan.

One basic reason for planning is to identify opportunities, such as opportunities to obtain more resources, redirect existing resources, hire new faculty, and influence individual faculty as they plan their own programs. Planning allows the department to go out and make the opportunities happen, rather than waiting for them to come knocking. The planning process needed to deal with the challenges facing land grant universities and colleges of agriculture should be decentralized and open. Based on the premise that faculty are the ultimate source of creativity (Beattie, 1983), the process should work from the faculty level up to the department level, rather than the other way around. Also, the plan should not become a straightjacket that stifles creativity. It should be a continuing process that encourages the infusion of new ideas. Print plans on recyclable paper, not as glossy publications. The greatest benefit often comes through participation in the planning process itself rather than from the product.

The planning process begins with a study of priorities but should include a strategy of continually broadening the circle of people involved in the planning process. Broadening this sphere of involvement includes appropriate composition of the steering committee, retreats, or workshops to ensure participation and feasibility. An issue to be resolved in developing this process is the degree of faculty involvement.

Faculty Involvement

The key issue in managing change is the degree of faculty involvement in planning and decision making. Although some may believe that faculty involvement should be maximized, not all faculty members can be, should be, or want to be involved in every decision made. The following three questions suggest the criteria to be used in determining the degree to which faculty should be involved in the decision-making process (Tucker, 1984):

1. Expertise. Who knows how to solve the problem? The administrator? A particular faculty member or group? The department?
2. Acceptance. Is faculty acceptance of the decision crucial for effective implementation? Will implementation fail if some or all of the faculty members refuse to go along with the decision?
3. Time. Is there enough time to get the faculty involved in the process of decision making? Is an immediate decision crucial?

These three criteria will vary in importance depending on the circumstances. Considering these criteria, however, will help the department head or faculty planning committee decide the extent to which faculty should be involved in the decision-making process.

Curricular change should be the product of individual and collective faculty thought and debate. Harl (1993) cautions that "ideas floated by an administration eager to capture the latest educational fad that are not subjected to the annealing heat of faculty debate are often doomed to failure or worse—misleading or misguiding a generation of students." There is still the chance that students will occasionally be misled or misguided, but the probabilities are much lower with active faculty involvement.

Universities, as compared with other social institutions, have a long tradition of collegiality. The idea of a community of scholars is compatible with the processes of shared decision making by equals, which is an idea that is now starting to catch on in corporate and governmental organizations. Collegiality or faculty involvement is a practice that should not be sacrificed to expediency.

Budget Considerations in Promoting Change

Significant educational change in a college or university frequently involves some redirection of resources (Mathews, 1990). Increasingly constrained budgets for higher education across the country, however, have reduced the ability of academic units to initiate change. Revising curricula, developing new courses, and adopting new teaching methods all require investments of time and money. The venture capital to invest in these changes is simply not available in the budgets of most academic units. After covering salaries and modest allocations for operating expenses, little is left to purchase equipment, books, and materials or to upgrade course content and teaching methods.

The same constraints apply when it comes to faculty time. With rigid accountability standards for teaching loads and expectations for "scholarly" research, faculty simply do not have the time to invest in upgrading course content, writing new textbooks, or developing presentation graphics to enrich their lectures. Furthermore, the development of more cohesive curricula requires a considerable time commitment working with colleagues both within and outside the department to coordinate course offerings and content.

The enthusiasm for change is quickly lost when faculty realize that the necessary resources will not be available to implement them. Like the traditional farmer, who says he already knows how to farm better than he is doing, faculty members respond that they already know what they should be doing to upgrade their current courses, develop new courses, and apply new teaching methods. The problem is having the time and resources to get the job done.

The question facing administrators is how to set aside part of the budget to invest in change. Coming up with these funds will require some hard decisions about not filling positions, reducing staff support, or discontinuing programs that have been in place for many years. Then, once the administrator has the funds in hand, the next question is where to invest it. Obviously, this must be done as a part of a long-range planning effort, with sufficient amounts of money allocated to complete the job. Seed money grants have not worked in teaching as they do for research because the sources of funds are generally not available to bring the change to fruition. Sound fiscal management practices need to be employed, however. Practices such as progress reports, evaluation of accomplishments, and reporting of results should be employed to ensure that the investment decision pays off as expected. The same issues apply to the management of faculty time. Administrators need to find ways to free faculty from the daily demands of meeting classes and attending committee meetings so that they can invest time in advancing the teaching program.

The capital stock of course work, teaching materials, and instructional methods in colleges of agriculture (and other higher education units) depreciates with time. Continuing investment is needed to keep it up to date and to move it ahead. Administrators must place a high priority on finding the time and money needed for this investment.

Strategies for Effecting Change

If it is accepted that changes are needed in the teaching programs in the colleges of agriculture, how do administrators go about promoting that change? According to Beattie's (1983) second commandment, "Always remember that everything relevant and beneficial . . . occurs at the most micro-level because of the imagination, creativity, drive/desire, and intellectual horsepower of the individual faculty member." If all creativity originates with the faculty, then the question becomes How do we encourage faculty to make these changes? Here are some ideas to get started:

1. Recognize the trends. Provide information to faculty regarding changes taking place outside the university that suggest the need for change. The administrator must determine the obstacles (Hefferlin, 1972).
2. Engage in planning. Involve the faculty in setting goals and deciding on the strategies for change. Build on existing concerns (Hefferlin, 1972).
3. Hire the right faculty. Two of the Beattie's (1983) commandments are relevant: "strive always to hire the best human talent possible, for it is that which is scarce" and "make the difficult negative decision on granting of tenure."
4. Provide incentives, if nothing more than a pat on the back. Develop reward systems to encourage teamwork and promote innovation.
5. Restructure budgets to provide flexibility for developing programs and offering new courses. Find the venture capital needed to develop and maintain new teaching methods and update knowledge.
6. Promote professional development—in the appropriate direction. Build the capabilities of faculty members to decide on positive changes and implement them.
7. Enhance organizational flexibility by breaking down the barriers between units within the college and between the college and other units to encourage cooperation. Facilitate the faculty member's desire to seek cooperation from the other college. Promote integration of teaching, research, and extension.
8. Reduce risk. If the risk to the individual from undertaking change cannot be reduced, it should be explained so at least faculty members enter into the situation informed. If you can't control the risk and protect the innovative faculty member, at least disclose it.
9. Although some programs must be discontinued, remember to respect the past (Hefferlin, 1972). Holt expresses this more pointedly: "Hold an official burial ceremony for the old program."

When we examine the reasons why successful academic programs endure and new innovative approaches are developed, we often find a champion who was personally committed to making it happen. A key to managing change in academic programs is to authorize and empower "change champions" among the faculty. Within the faculty, certain individuals are change oriented. They are risk takers, energetic, dedicated to teaching, and knowledgeable. These change champions should be involved in the process from the beginning. Those selected to be change champions should be respected by their peers, sensitive to campus politics, and willing to help make the difficult choices (Wood, 1990). These champions should be doers who both plan and carry out the agenda, not a committee that writes a report calling for someone else to do something (Farmer, 1990, p. 13). Identifying and commissioning these faculty leaders requires administrators who know their faculty and who are willing to delegate control of the process. The challenge to the administrator is to empower them to proceed.

Although consensus management (where an eleven to one vote represents a tie) has many positive attributes, it may not foster the dramatic change required in many institutions. Administrators are well advised to work with and through faculty members and staff, but bold initiatives are seldom born in committees (Rush, 1992).

The Need for Leadership

"Academic leaders who desire change or see its need must create environments that encourage innovative thinking and risk taking," writes Farmer (1990). The decision to change comes with leadership. He elaborates that as change agents these leaders play several roles:

- catalyst, helping others understand the need for change;
- solution giver, offering specific implementation strategies;
- process helper, helping others in developing successful innovations;
- resource linker, bringing resources together to advance the change;
- confidence builder, promoting a sense of stability and possibility.

The key to the basic change required in higher education is leadership.

Institutional Impediments to Change

Many obstacles must be considered if curricular change is to be implemented. It is imperative that these obstacles be analyzed and efforts be made to understand and overcome them. All discussions on systemic change will be moot if there is failure to facilitate and implement these changes.

Kanter (1992) concluded that institutions are more skillful at designing change than at executing it and that a better implementation process is necessary to bring about change. Change is often associated with resistance and fear, and in most situations incremental change is both more acceptable and less discomforting. Systemic change implies structural rather than procedural/cosmetic change and consequently is likely to elicit a much greater degree of resistance than incremental change. Again, it is to be noted that systemic change rather than a piecemeal response is necessary if colleges of agriculture are to respond to the challenges they face.

Change will be facilitated if there is rationale and justification for the change. The change must be perceived as an improvement in the quality of education. Faculty involvement and an open discussion of desired change must be based on the existing perception of the current educational shortcoming of colleges of agriculture for the education required for the twenty-first century.

Faculty Interests

Meyer (1993) used the Delphi technique with sixty-two informed respondents to identify the issues and challenges affecting the future of "colleges whose historical roots were in agriculture." One of the important challenges identified is the need to improve faculty interest in undergraduates. The respondents felt that faculty members are not as interested as they should be in undergraduate teaching because of its incompatibility with their interest in research. This may be reinforced and perhaps fostered by the reward system within the university. Meyer holds the proposition that faculty interests and breadth of scholarship need to be expanded so that they can deliver comprehensive and timely undergraduate programs. Lack of faculty interest in undergraduate education must be overcome before change can occur in undergraduate education.

Organizational Structures

Organization theory holds that it is important to remember that most organizations value stability and predictability (Turner, 1990) and consequently do not readily embrace change. This is especially true of structured, departmentalized, hierarchical, and tradition-bound colleges. A rigid hierarchical organizational structure discourages experimentation and creativity. Thus, it is important that organizational structure be streamlined and faculty participation in decision making be increased if systemic change is to be brought about.

The organizational structures of colleges of agriculture typically vest strong power bases within academic departments (Meyer, 1993). These departmental boundaries may discourage cooperation among individual faculty members, allow duplication of effort, constrain curriculum coordination across units, and deter development of interdisciplinary teaching programs. Cooperation and the integration of course work across disciplines is necessary to successfully implement systemic change.

In spite of these disadvantages, the departmentalized structure has several important functions.

The departmentalized structure is not necessarily outmoded. Through their departments, faculty members share responsibilities for (1) program development and planning, (2) resource allocation, (3) professional development, and (4) quality control. Academic departments, however, face a significant challenge in balancing their roles as keepers of the discipline and contributors to the profession on one hand and as responsible providers of education and research in support of the land grant mission on the other. The challenge is to turn this potential conflict into creative tension. To accomplish this, faculty members have a responsibility to work (1) within the disciplines and their professional associations to assure that they are responsive to societal needs and are open to new ideas and approaches and (2) within the land grant system and with its clientele to help them understand the importance of disciplinary standards and peer review.

Institutional Culture

Organizational culture refers to "the way we do things around here" and can be a significant deterrent to change (Newton and Tarrant, 1982). Some institutional cultures are more conducive to change than others. According to Hefferlin (1972), each college and department has its own historic orientation toward change. This institutional ethos toward change is likely to be self-perpetuating because faculties choose administrators similar to themselves to lead them and hire new faculty with similar dispositions. Because it is easier to replace persons than to change their attitudes, changes are more likely to occur when old professors retire and new ones are hired or when current administrators are replaced by others. Effective leadership and collaborative planning, if they are part of the organizational culture, can be positive forces for change. Implementation of change will likely be a matter of developing a culture for change in the institution.

The Image of Agriculture

Agriculture has been losing its uniqueness. As agriculture loses its uniqueness, public support for these institutions that serve a diminishing number of people decreases. Agricultural colleges and agricultural courses are also losing much of their uniqueness. Less emphasis is being given to agricultural examples and applications. The reasons for this tendency include fewer faculty and students with agricultural backgrounds, the decrease in availability of current agricultural textbooks and teaching materials, and the desire for a general ("science based") curriculum to prepare students for the likelihood of career changes and employment in nonagricultural occupations.

Unless a unique mission of teaching programs in agricultural colleges is emphasized, it will be difficult to muster support for change. Without change to distinguish the curricula offered by colleges of agriculture from the curricula of other colleges, the result can be the gradual absorption of agricultural programs by their parent disciplines. The perceptions that the curricula offered by colleges of agriculture are solely concerned with crop production and animal husbandry are detrimental to recruiting students and to competing for limited funds within the university. Agricultural colleges face a major marketing challenge in telling the story about the opportunities for exciting work in agribusiness, international commerce, biotechnology, human nutrition, food distribution, natural resource management, and environmental remediation, in essence, the careers related to biological systems in support of human societies.

Academic Freedom

Within academic circles, curricula are generally a corporate concern, but there is hesitance to question course content. That domain is left to the instructor. This may be in deference to academic freedom, but as a result curricula may not be internally coordinated and the bits and pieces of subject matter not well integrated from the student's perspective. Thus, academic freedom can be a significant impediment to change within higher agricultural education. Harl (1993) points out that individual faculty members may thwart the reform process by continuing to teach the way they have been teaching. Just because a department head or group of colleagues suggests that a course should be taught differently does not mean that it will be. In fact, the actual content of the course may not be known other than only very generally by faculty colleagues because of the lack of peer review and administrative oversight.

Manderscheid (1988) spoke to the challenges of designing a relevant curriculum: "Indeed, one of the difficulties is that many teaching faculty think only of 'my course' rather than about an undergraduate education." Although the curriculum design should consider the strengths of the current faculty, these people will not always be available. Thus, it is important to separate the courses from the persons who are currently teaching them and to consider the difficult decision to eliminate someone's course if that is what is needed to build a stronger program with limited resources. The ultimate objective should be not to teach undergraduates what to think, but how to think.

Lack of Knowledge

In order to promote and participate in the process of change, faculty members must know why the change is needed and have the expertise to make the needed changes in their teaching programs (Harl, 1993).

The problem, however, may not be the lack of specialized knowledge. According to Meyer (1993), the problem is a shortage of rigorous generalists among the faculty to respond to the need to cover more contemporary issues, especially those related to multidisciplinary problem-solving approaches. The generalist here is one who has been able to move competently from specialty to specialty (Drucker, 1994). The current reward system and the competition of research contribute to this shortage of a wider knowledge base to support the revision of teaching programs.

Managing the Stress

Change will bring stress. Helping people to manage this stress is an important function of the administrator. Stress in organizations undergoing change is inevitable but should not be ignored. The sources of stress need to be recognized and the effects understood. Managing stress involves (1) identifying the sources of stress, (2) examining the effects of this stress, and (3) reducing this stress through listening, support, counseling, and so forth.

Humans probably perform best when they have some stress, but the problem is to find the right amount. Too little stress results in underachievement, and too much results in frustration and conflict.

The standard techniques of behavior modification can be used for managing the stress associated with change. These techniques involve agreeing and insisting on very small but progressive steps forward while providing personal support and praise for progress.

Motivating the Faculty

Incentives influence what people do. "Incentives matter—in universities as well as in the 'real world,' " according to Beattie (1983). But what are these incentives? What motivates faculty? According to Castle (1991), experience has shown that at least three factors are important in bringing forth change within the academic environment:

1. Funding. Academics respond to economic incentives. Unfortunately, the budgets for teaching programs within agricultural colleges have become more limited, making it difficult to provide these economic incentives, but there are other incentives.
2. Need. Faculty members, much like other people, respond to felt needs. Motivating change, thus, involves providing evidence of the need. Motivating change in teaching programs, for example, might involve providing more information about the characteristics and goals of the students being taught and the expectations of their future employers.
3. Intellectual excitement. Intellectual effort is often stimulated by conceptual breakthroughs and discoveries and the existence of unexplained anomalies. Thus, another way to motivate change in academic programs in agriculture is to provide a conceptual basis for the changes. These concepts should then stimulate the intellectual work needed to apply the ideas to revise curricula and develop new teaching methods.

Elaborating on this last point, it is the resolution of contradiction that often leads to major breakthroughs. Managers, administrators, and institutions must be able to consider opposing views. Integrative scholarship can capture and summarize the intuitions that we have often come to when we think about a problem (Kunkel, 1997). It can help clarify thoughts because, if appropriately approached, it orders and focuses a wide range of different intuitions (Elliott, 1992).

Tucker (1984) suggests three additional incentives that may be important in motivating faculty members to change:

1. Professional advancement. Faculty members are motivated by the desire for competence and the need to achieve. Therefore, they are willing to participate in planning and implementing change for the sake of their own professional advancement and the department's or college's improvement.
2. Compensation. Some faculty are motivated to change only if they can anticipate a more tangible reward that they can share with their families. In other words, they vote their pocketbook. Failure to receive compensation will lead faculty members to seek other employment or to retreat into their own work. Either way, change may not be served if the faculty member is not replaced.
3. Negative consequences. Some faculty members are motivated to change only when they consider the consequences of not changing. For example, a department with decreasing enrollments will be more easily motivated to change its recruiting policies and curriculum if decreasing enrollment is expected to lead to a reduced departmental budget and fewer faculty positions.

Developing the Implementation Processes

Many faculty members are open to change, understandably some more than others. However, they must be provided assistance in curricular revision, especially in recognizing what is being planned and its relevance and value. Ideally, change in curricula ultimately should be a bottom-up process requiring faculty involvement and ownership. This will require that faculty subscribe to the institutional goals and be committed toward working for these goals. However, simply subscribing to the goals is not going to bring about change. Colleges of agriculture wishing to pursue modernization of their curricula will likely start small and build. A few key instructors (not courses), those committed to quality teaching, should be identified for the development and implementation of new forms of higher education and be compensated for their effort and willingness to develop such courses. After the instructors have made changes appropriate to concerns within their college's curricular strategy and have taught their respective courses, case studies of the successes and failures can be presented to the faculty as a whole. This will take time but is more likely to be successful than a strategy that favors rapid changes without careful forethought. Change will require support and encouragement ranging from released time for faculty to attend meetings, funds for travel to professional meetings, and funds for invited educators to present workshops on campus on the direction of the change and how to bring this about. This likely will be a slow process requiring fairly lengthy deliberation. Thus, senior faculty and administrators must become intellectual leaders, provide encouragement, and have a great deal of patience.

Also, higher education in agriculture must seek assistance and involvement from other colleges on campus. Individuals in colleges of education can be invaluable in assisting with curricular change; similarly, interaction with individuals in colleges of liberal arts and business sciences will allow a multidisciplinary examination of our courses. It is important to seek ways to eliminate artificial boundaries and barriers that exist on campuses (Ward, 1994). Further, we must avoid being bound by tradition, and in addition to agribusiness interests we must include environmentalists and other groups that have an interest in agricultural systems in deliberations relating to curricular change.

Conclusion

Changes in direction will not happen overnight (Newton and Tarrant, 1992). The process will be fraught with problems, pain, contradictions, and new challenges. Much will be asked of leaders and managers today and tomorrow if the transition into the twenty-first century is to be educationally smooth and prosperous. To assure the education and training needed to meet the world's future food and fiber needs, the task must be undertaken. Despite, however, the urgency and priority of these changes, such undertakings do involve risk. More than one has lost his or her job in challenging institutional culture and inertia (the status quo).

The most important task of a leader is creating the climate that is conducive to change. The development of orderly problem-solving processes will help. This includes developing procedures through which faculty can participate in the decision-making process. Faculty members who are affected by change should understand the change and its consequences. In any planned change the leader must give as much attention to the emotional aspects as is given to the information aspects of the change. This is discussed further in the next chapter.

Deciding that change is necessary is the first step in improving a process or a curriculum, but it is not enough by itself. Understanding the process of change and what needs to be done to institutionalize the change and to make it last is necessary if the desired outcome is ever to be achieved (Curry, 1992).

Both the faculty and administration are key to implementation. Administrators can perhaps serve the role of visioning or catalyzing the vision and subsequently be involved in facilitating change. Administrators can plant ideas with selected faculty members, faculty members who are well respected within academic units, and then back away and give the faculty the responsibility for the follow-through. It is important for administrators to be supportive of the faculty members in their task, so that faculty members have a sense of security, if they are concerned with undergraduate education. It is critical that heads of department, associate deans, deans, and directors support curricular change.

Administrators are strategically placed so that they are the ones who can effect change across departments and other colleges. Changes have occurred at institutions as the dean of agriculture works with deans of other colleges. Cooperation, jointly working together, has developed programs in agribusiness, agricultural ethics, biotechnology, bioremediation, nutrition, and others.

Faculty involvement is a key component of managing change. Faculty, however, will each respond differently to change. Successfully introducing change requires that administrators know their faculty members individually and understand why they respond differently. Developing appropriate incentives and motivation techniques will help offset the normal resistance to change and involve faculty in the process.

6

Summary and Synthesis of Principles for Systemic Change

The Work Group

The focus thus far, on fundamental change in higher education in agriculture and natural resources, has examined the theoretical and methodological issues concerning needed change. The most significant conclusion is that all colleges must rethink their values, missions, and goals, complementary to a changing society and the new world views of food and to biological and ecological systems. The key premise is that colleges of agriculture and their related and derived units are the academic structures that should provide the educational, scientific, and scholarly framework for the understanding, development, management, and use of the world's biologically and ecologically based systems and their relevant human resource systems for the benefit of human and natural societies. Thus, they are counterpoints of engineering and business colleges that are concerned with physical and financial systems for human societies.

Specific recommendations for the future are as follows:

1. Renewal and rethinking of the academic mandate of higher education in agriculture must be based on principles congruent with the needs of society in the twenty-first century, the substantial transformation proceeding in the global food and resource system, and the significant and continuing accumulation of scientific expertise and knowledge.
2. The knowledge base for higher education is expanding exponentially, providing a science-laden education requiring development of problem-solving and communications skills of undergraduates for employment and advancement in the agriculture industry; thus, the integration across traditional disciplinary areas must occur.
3. Higher education must redefine the values upon which reform will occur to include beliefs about what is important, the need for individual action, appropriate infrastructure, and procedures that will guide colleges' and universities' activities.

4. Higher education must adopt a vision for excellence that will lead to redesign of curriculum and contemporary pedagogy.
5. Higher education for the twenty-first century must be based on needs of students. Colleges must develop life-long learning paradigms to address on-campus students and the needs of alumni.
6. As the professions in agriculture and life sciences are redefined, new models for higher education must be created. Implementation of change in higher education should be built on a system that values excellence and integration across teaching, extension, and research.
7. Systemic change will lead to significant change in the involvement of students, administrators, faculty, employers, and other key individuals for developing the mission, goals, and curricula in higher education.

Project Interact (Kunkel, 1992) and the NRC-USDA–sponsored national conference in 1991 (NRC, 1992) called for systemic, not just incremental, change in undergraduate education in agriculture. These efforts also sensed a critical framework: the system is being forced to be leaner at a time when a greater diversity of students is seeking education in a larger range of subject matter. At issue is the connectivity of the land grant institution and higher education in agriculture to the changing society.

It is likely that market economics of delivering education will define the usefulness of higher education in agriculture. The market may say that any university degree program will do, that education (such as in agriculture) must be specialized and in depth to be useful, or that the education must be directed toward a different market than it is now. What has been missing is a comprehensive guidance that can encourage the needed changes. Part 1 of this book is a considered attempt to review the theoretical bases of higher education in agriculture and provide the framework and principles for systemic change, not incremental change alone.

Whether change is a leap or an evolution, which is the more likely situation, the distance that change moves the program from the current situation to meeting the needs of education in the twenty-first century is a critical factor. The incentive for change must be present; the best incentives for change can be created by the reaching for understanding of what ought to characterize scholarship in a college of agriculture.

The call for systemic change in agriculture curricula is more than just a restructuring of educational goals and an attempt to regain the traditional status of land grant education. It is through change that excellence in education can be achieved. It is a common belief that excellence is related to quantity of resources: funding per student, faculty-student ratios, facilities, faculty salaries, and so on. Excellence, however, is more directly related to vision and the concentration of people's energies through establishing priorities and a commonness of goals. Excellence does not start with money. Today's and tomorrow's challenge is to provide quality with limited resources.

The intellectual purposes of the modern colleges of agriculture are becoming divergent. The unit that has been called the college of agriculture, food, and/or natural resources may increasingly provide the biological, physical, and human dimensions of issues related to ecosystems. Some colleges, more traditional in their world views, may focus on the entire agribusiness community and the agricultural industry. Still others see their role in developing students' professional and technical knowledge but also in providing for the growth of students in their abilities to resolve a range of societal and technical problems, their skills in communication and relationships to others, and their sense of values. The focus should remain in areas of food and resource management, but higher education in agriculture can be defined much more broadly in the future.

Colleges of agriculture and related and contemporary derivatives may be redefined as the academic structures that provide the educational, scientific, and scholarly framework for the understanding, management, and use of biological and ecological systems and the relevant human resource systems for the benefit of human and natural societies and their communities.

Two primary tenets seem to emerge for higher education in agriculture:

First, the clear requirement of colleges of agriculture in the future is to (re-) establish the connectivity between the higher education system and the society it serves. In part, that means sensing where agriculture must focus in the future—feeding a growing world population, the increasing pressure on resources, reconfiguration of the economic and political power of the food and natural resource systems, the industrialization of agriculture and its constraints in the future, the changing rural environments, and a growing plurality of views about food and agriculture—and designing educational programs accordingly.

Second, the college of agriculture should put the student first. Programs should be tailored to meet student needs, not the convenience of faculty and staff. How the student is viewed may be the most effective systemic change that can be made. Do we envision education as producing leaders, scholars, technicians, or practitioners or those who are employable in a particular segment of agriculture, food, and natural resources or what?

An initial set of concepts for systemic change in higher education has been drawn from focus group discussions. These concepts can be stated as goals for change; although, as we draw them, they are working principles subject to debate, refinement, and expansion. These principles are listed below:

1. Define agriculture broadly. Agriculture must be recognized as a system based upon the stewardship of resources and the environment,

maintenance of the public health, and the livelihoods of billions of people in the world. Agriculture today includes, but is not limited to, the elements of biology and ecology; land, air, and water resources; production processes; marketing; transformation and use of food; consumption and human health; recycling; and waste disposal. Agriculture thus depends on the social, engineering, and physical sciences, as well as the biological, and encompasses the economic, marketing, and financial systems that bring these biological, resource, engineering, and physical systems together and make them operate (Pulver, 1994). The goals and values of people working on every level of the food, agriculture, and natural resource system require consideration in higher education in agriculture. (See chapter 2, "Changing Knowledge Base" section.)

2. Begin with new visions of the educational roles of colleges of agriculture. Colleges of agriculture developed with a vision about how things should be. The principal and initial point of departure of an institution toward systemic reform is the determination of the vision for the future. Vision, in this context, refers to what the institution believes it ought to be about and whom it should serve. The vision provides a sense of direction for the change process. The vision, in large measure, determines the strategies and goals of the educational program. But if the institution is to deal with the theoretical bases for systemic change, it will also need to deal with its values. Both the vision and the perceived mission of the unit are derived from the institutional values and faculty beliefs. Any effort toward systemic change must, therefore, involve the faculty of the institution in a process of articulating existing values and critically evaluating their contribution to the vision of the college. (See chapter 3, "Envisioning Excellence and Change" section.)
3. Accept broader responsibility for education in the university. Colleges of agriculture must develop closer linkages and collaborative efforts in both teaching and research with other units of their universities. Simultaneously, they need to address the general educational needs of all students in the university in terms of food and nutrition, resource use and conservation, environmental degradation, and ecosystem sustainability. (See chapter 3, "The Image" section.)
4. Set a high priority for undergraduate education. The foremost basis of systemic change in higher education in agriculture is that the college or department focuses on a limited number of priorities, making sure that one of them is undergraduate education (Magrath, 1994a).

5. Target recruitment efforts to the diversity. Colleges of agriculture must now position themselves to attract and recruit qualified students as individuals. Hispanic, African, Native, and Asian Americans, students with children, physically handicapped students, nontraditional students, and students from rural areas and inner cities all present different challenges to the system of higher education. (See chapter 3, "Putting the Student First" section.)
6. Academic needs instead of research needs should drive strategic planning. During recent decades, the research initiative, either faculty or institutionally based, has driven the academic program. External factors—available grant funding, political power, clientele interests, public policy—may determine what research faculty and, hence, what teaching faculty will be hired or retained. Adjusting curricula to the changing future needs of the students requires equal consideration be given to the academic component in strategic planning within the institutions. Planning cannot be done in an academic vacuum. (See chapter 5.)
7. Develop partnerships. Increasingly, it is evident that institutions must collaborate with other institutions, in the state, region, nation, and world. Inter- and intra-institutional collaboration is one way to deal with the need to set priorities, but it requires clear external communication and internal support. (See chapter 3, "Collaborations" section.)
8. Develop different programs for different institutions. The nearly common missions and course offerings of undergraduate education in traditional colleges of agriculture are likely a matter of the past, not the future. But the special qualities of colleges of agriculture remain an important resource that should be preserved and applied to both continuing and new educational ventures. As they do so, niches of programmatic specialization will evolve and unwarranted duplication will be reduced. These changes should not be necessarily considered a matter of retrenchment. Rather, the specialization and differentiation of programmatic efforts between and among institutions should be consonant with societal needs, the needs of constituents, and the new challenges and opportunities within the university and college providing higher education in agriculture. (See education discussed in chapter 3.)
9. Understand the knowledge base. The knowledge bases of agriculture need definition in each institution. For example, to the extent that traditional (artisanal) agriculture is to be considered, the operation of a complex system would be the focus. Persistence (sustainability) of

the system is a component. If increasingly industrialized agriculture is the focus, specialized knowledge may be necessary. The educational approaches may be different for each focus, but understanding both complex farming and the industrialized food system may be necessary to provide education for the flexibility of graduates in the future. (See chapter 2, "Changing Knowledge Base" section.)

10. Emphasize the distinctive role in the concept of production, but not exclusively. Distinctively, colleges of agriculture understand food, fiber, and timber production and can and should continue to take leadership roles in the productive uses of the biological systems and natural resources. The question of production is unlikely to come up in a college of liberal arts or a college of science or an environmental science program. Colleges of agriculture, however, have often focused too much on production to the exclusion of other needed facets of related education for the future. (See chapter 2, "Intellectual Purpose of the College of Agriculture" section.)
11. Consider values and how they change. Historically, a set of values has been promulgated by colleges of agriculture. These values have been inherent in marketing the college and its programs and courses. Students entering colleges of agriculture also have been a self-selected group, often motivated by an affection for plants and animals. Others come to the college of agriculture because they feel comfortable with curricular bridging disciplines. Others relate to societal and business aspects. And others find the science programs inviting. These historic values of the college may not be a distinctive feature of colleges of agriculture in the future. Change, therefore, should accommodate values of a changing economy and society. (See changing values discussed in chapter 1.)
12. Integrate the program. Change should correct discontinuities of higher education in agriculture. The revitalized program will have to be integrative. It should include people who understand economics, human values and culture as well as the disciplines of agricultural science, nutrition, and the natural resources. Integration is more than being multidisciplinary; it is a frame of mind, a thought process, alien to reductive thought. It may be the only way colleges of agriculture can meet their mission of providing education of excellence. Global dimensions should penetrate the curriculum, not be limited to a few specialized courses. The curriculum content should consider both the international implications of trade and the relationship to local agriculture and cultures in developing countries as well as in developed countries. The creation of greater awareness and understanding of

local and global environmental concerns will be needed for graduates in the twenty-first century. Similarly, understanding of contemporary issues will be needed to build the societal readiness of graduates in agriculture and natural resources. (See chapter 2, "Changing Context" section.)

13. Utilize teamwork and systems thinking. Increased emphasis on intellectual teamwork and use of systems approaches is essential to a viable future for higher education in agriculture. The scope and complexity of agriculture and related fields have become far too great for mastery by single minds (no matter how nimble) working in isolation. Agriculturalists of the future increasingly will need to integrate their individual experiences and talents to achieve a coordinated, cooperative effort for understanding, communicating, and solving problems. Appropriate academic preparation must include a stronger integrative foundation, which intellectual teamwork in a systems context can provide. (See chapter 4.)
14. Organize cores of study and add perspectives. Organization of the modern undergraduate agricultural program calls for the development of cores of study: a foundational core, a functional core, and an integrative core. Added to these cores ought to be a set of perspective courses and courses for leadership: global issues, political and regulatory systems, intellectual property, contemporary issues including social concerns, and human values. (See chapter 4.)
15. Diminish redundancy and enhance flexibility. When integrated learning becomes a standard part of the course objective, redundancy will diminish and opportunities for flexibility will be enhanced. A certain redundancy occurs because each of us teaches our courses as an end to itself or so that students may take the next higher course that we teach. Redundancy, thus, places an emphasis on course content. Flexibility places an emphasis on individual teaching capabilities and individualized courses of study. Less emphasis should be placed on what students are learning; more emphasis can then be placed on the quality and productivity of learning. (See chapter 3.)
16. Connect the courses, but with care. The principal deficiency of higher education, as perceived by students, in agriculture and in general, is the lack of connection among the several courses in the curriculum. Knowledge taught for its own sake often loses its relevance to the capacity for generalization. Thus, the basic needs of colleges for teaching competence and also motivating students to careers in agricultural sciences may not be well served by the traditional programs. Coherence for its own sake, however, may negate its positive

values by making it more difficult for students to sample other disciplines. (See chapter 2, "Changing Knowledge Base" section.)

17. Blend the teaching of knowledge and skills. Theory in higher education suggests the need for a concept of blending knowledge and skills to achieve the expected outcomes of the undergraduate program. The studies of processes and of such qualities as the abilities to write, speak, listen, and work in teams, problem-solving abilities, and leadership should be integrated into curricula designed for technical competence. These skills rank high among employers as important characteristics of the college graduate. Improvement of methods—modules, internships, enhanced semesters, inclusion of current issues—should be sought to provide flexibility, efficiency, and enrichment of education. (See chapter 4.)
18. Expand the clientele base. The future of colleges of agriculture and natural resources is more closely related to their understanding and responsiveness to the external environment than for any other unit of a university. Colleges must endeavor to expand their clientele to include a broad array of consumer, environmental, conservation, and animal protection groups. It will be difficult to address the issues raised by the new groups while maintaining positive relations with commodity groups. However, to ignore these new groups would reduce the political power base of agriculture and be contrary to the directions of resource conservation and stewardship that agriculture espouses. (See chapter 1, "Changes in Clientele" section.)
19. Teach students how to think, not what to think. Students should be taught how to gather facts and do the in-depth thinking required for rational, philosophically consistent decision making. They should be taught the limits of science. The colleges should provide a diversity of knowledge without bias and teach the students how to come to a conclusion but not force the conclusion itself. (See chapter 3.)
20. Try some new methods of teaching. The implementation of systemic change within the college of agriculture should use the best educational methods and tools available for each specific task. Traditional lecture, team-taught, modular, and capstone courses should be used as appropriate, based on the particular learning objectives, but should make use of the new technologies enabling greater student participation. Tools such as case studies, experiential learning, collaborative learning, distance learning, digital technology, internships, and class discussion must also be chosen with forethought to achieve the desired learning goal as effectively as possible. Faculty will need to use the new models of instruction. (See chapter 3.)

21. Do not simply say it, do it. Enabling systemic change is crucial. Institutions must not simply focus on the design of change but must also develop the implementation process to bring about the change. New ideas will be useful. Adequate faculty start-up funding for teaching, analogous to the research start-up, is a novel idea whose time has come. Adequate continuing funding is also needed. There must be opportunity for experimentation in academic instruction. Facilitating change will require a combination of intuition, good sense, and courage.
22. If it is done, reward it. Ultimately, it is imperative that an administrative structure and a reward/recognition system be adopted that support, rather than inhibit, the use of innovative educational methods and tools. Appropriate reward and recognition systems must be in place, for without them, even with time and money available, few faculty will bother to follow through. (See chapter 5.)

Part Two

The Focus on Implementation and Agenda for Action

Among several studies of higher education in agriculture, food, natural resources, and life sciences, largely in the land grant universities, this was a study, first, of the theoretical bases for systemic change in the education and, second, for strategies for implementing the needed changes. The first phase of the work was published in 1996 (Kunkel et al., 1996) and is revised in part 1 of this book. The results of the second phase are the subject of this part. Together they make a coherent whole of higher education in agriculture and its component areas.

The initial grant was a USDA Higher Education Programs Challenge Grant initiated 1 September 1993. The second grant was entitled Strategies for Implementation of Systemic Changes in Higher Education in Agriculture. A smaller core work group was involved in the implementation phase, but the same institutions as in the first phase—Cornell University, University of Connecticut, Alabama A&M University, Rutgers University, and Texas A&M University—were represented.

The implementation study was exploratory. The initial phase of the implementation phase involved presentations at the annual regional teaching symposia held under the auspices of the Academic Programs Section of the National Association of State Universities and Land Grant Colleges. The presentations were well received by the participants, but little evidence was obtained that such a process would lead to any fundamental change in the educational programs of institutions. The participants in the regional teaching symposia were teachers dedicated to the improvement of their own skills. This, of course, is an admirable and needed attribute, but it did not seem to be a strong incentive for involvement in institutional change.

Blame, however, should not be placed on the participants. The process of change, if it is to be fundamental, requires a search for penetrating ideas. Also, the principles of change had not been distilled to the level that they made sense to those who can contribute to the process of implementation. The need for another approach became obvious.

The search for solutions turned to workshops, or focus groups, in which concepts would be explored and participants could engage freely in discussion. Two workshops were held, one in May 1997 at Texas A&M University and a second one in April 1998 at Cornell University.

The procedure was this: Members of a core work group evaluated the bases for systemic change and selected certain approaches that might be keys to implementation of change. Plenary speakers were identified and were invited to present exploratory concepts. Participant discussions followed the presentations at the workshops. The approaches selected for the second workshop emerged out of the thoughts expressed in the first workshop. Both workshops, or focus groups proceedings, were recorded on video tape and subsequently transcribed.

Part 2 is a summary synthesis made out of the presentations and discussions at the two workshops. The ideas expressed in these workshops were multiple and successively creative. Proceedings in the usual format of submitted manuscripts would not likely provide the important features of the creative process that occurred. Thus, both the plenary presentations and the participant discussions are summarized and brought into a focus on the implementation of change.

As the rethinking of goals and missions of colleges of agriculture and natural resources is accomplished, implementation of change will require an understanding of the framework of the institution and of what it must do to respond. The introduction of new concepts of what higher education in agriculture should be and do cannot alone bring about the needed fundamental realignments. Nor can organizational and managerial changes do this alone. To have impact, the efforts to implement change must be coupled with changes in both the decision-making processes and institutional structures. It seemed important to revisit and analyze the theoretical structure and flow of one aspect of the college of agriculture, namely, undergraduate and professional education, as the basis for its reconstruction. The purpose was to engage in the process of discovery and development of applicable concepts that individual institutions might use to reach the academic potential envisioned by the thinkers engaged in reexamining the colleges of agriculture and related areas and to follow through with implementation.

We note the efforts of Sherry Kelly of Texas A&M University and H. Dean Sutphin and Suzanne Alexander of Cornell University in handling the arrangements at the workshops and of the students who were our ever-ready technicians.

7

Responsibilities and Expectations

Key Concepts

The participants in the workshop at Texas A&M University laid out several strategies for which administration, faculty, and clientele (students) should take responsibility. The significant feature of these strategies is that students as well as the administration and faculty have responsibilities, not just expectations.

Administrators

1. Administrators should document teaching responsibilities in position descriptions and reward achievement and innovation, giving the same weight as research.
2. Administrators should make the difficult decisions to provide the needed resources for excellence in all aspects of academic programs.
3. Administrators should establish continuous improvement teams to promote excellence in all aspects of academic programs, including participation in workshops and continuing education by both faculty and administrators.

Faculty

1. Faculty should value teaching, advising, and the scholarship of teaching and demonstrate enthusiasm for these important faculty roles.
2. Faculty should participate with other faculty in study and discourse about teaching, learning, and curricular design with the land grant mission.
3. Faculty should utilize integration of knowledge and systems thinking and insertion of values into the learning experience.
4. Faculty should emphasize the development of teaching skills in the training of new entrants to academic professions.

5. Faculty should experiment with and develop new teaching techniques to address diverse learning styles.

Students

1. Students should become motivated to take responsibility for their own learning (to view the university as a pasture where they graze for knowledge).
2. Students should engage themselves in designing their own undergraduate programs, taking advantage of opportunities for experiential learning and leadership development.
3. Students should participate as active members of a learning community of faculty, students, and stakeholders.
4. Students should take responsibility as students and alumni to provide feedback for continuing improvement of academic programs in the university as well as agriculture and natural resources.

Introduction

The sticking point in systemic change is the implementation. Incremental change is fairly easy to accomplish, but systemic change will not occur by simply exposing faculty members and administrations to books and conferences. What is required for implementation is debate, seeking the faculty with the knowledge base, networks, and writings that can become the engine of change.

As a result of the many forward discussions in the projects on theory and strategies for implementation of fundamental change, we now believe that the implementations of such change can occur if there is an appropriate conceptual basis that facilitates the change. The project has sought such conceptualization in two workshops focused on implementation. This and subsequent chapters are derived from the workshops at Texas A&M University and at Cornell University.

Separate Responsibilities

The guiding thoughts about the focus on implementation were that the introduction of certain themes in the course contents and curricular structures might enable the needed change. Thus, in the workshop at Texas A&M University, we tested systems thinking, values and ethics, and participatory education as possible avenues to fundamental change. But in the minds of the workshop participants, the responsibilities of administrators with respect to the teaching function, the importance that faculty be concerned with institutional change, and the obligations of the clientele (the students) to be concerned with their own learning emerged as the greatest needs in revitalization. There is little question that these components are important in facilitating the makeover that the colleges of agriculture, natural resources, and life sciences need for the twenty-first century. A principle concept involved that administrators and faculty have clear responsibilities in the strategies for implementation that may be shared but also can be clearly distinguished. Equally important is the notion that the clientele of the land grant university have their own responsibilities, be they as students or as other shareholders.

Shared Responsibilities: Initiatives in Three Colleges

Elements of shared responsibility between faculty and administration were enumerated by Professor A. Gene Nelson, Department of Agricultural Economics at Texas A&M University. The ways through which administrators serve as change agents are

- serve as catalysts helping to better understand the need for change;
- help find solutions by developing strategies for implementation;
- facilitate the process of change by developing strategies for implementation;
- help to bring together resources to get the job done (resource linkers); and
- build confidence by promoting a sense of stability and responsibility in the institution.

Three administrators—McArthur Floyd, associate dean for research, Agriculture and Environmental Sciences, Alabama A&M University; Suman Singha, associate dean, Agriculture and Natural Resources, University of Connecticut; and H. Dean Sutphin, associate dean, Agriculture and Life Sciences, Cornell University—provided reflections on facilitation of implementation at their respective colleges of agriculture.

Alabama A&M University (McArthur Floyd)

Floyd maintained that the 1890 land grant university, as a small institution, must define its own identity and not let others do that for it. Teaching is a historic role of the 1890 institution, but the other historic roles—research and extension—of the land grant university were added in 1972. The decision now is whether the 1890 institution is a teaching or a more complex, research university.

The task is to manage multiple missions. At Alabama A&M University, the title of research director was given up by the dean, who would focus his energies on education, with the intent of creating a student-centered, student-first environment. During employment of a new faculty member, the importance of undergraduate education is emphasized. During the interview process, the prospective faculty member is actually placed in a classroom setting to obtain feedback from students as to their ability to understand and connect with the prospective faculty member. The evaluation is then used in the employment process.

The College of Agriculture and Environmental Sciences at Alabama A&M University extends such emphasis to the existing faculty. Workshops and a task force deal with faculty rewards for teaching. Teaching portfolios are included in the evaluation system. Several external initiatives have provided opportunities for faculty members to improve their teaching. Two of the four departments and the Division of Family and Consumer Science in the college have undertaken curricular revision. There is an interest in reduction of redundancy in course offerings and incorporation of more modularization. The college is looking to increasing experiential, community-based learning by actual involvement of students in a community to solve "real world" problems. Such involvement includes nutrition, conservation, and land planning. Other goals are to increase the students' abilities to use system thinking and to infer ethical and international issues in the curriculum.

University of Connecticut (Suman Singha)

The issues at the University of Connecticut take on other dimensions. The College of Agriculture and Natural Resources works under a strategic plan for the university overseen by the Board of Trustees. New administrative structures at both the university and colleges encourage change. Here, too, the mission and goals must be described by the institution.

The new environment requires that more be done with less. Continual increases in student fees may price the institution out of the market. The competition to serve the general public appears to be from the community/technical college. Priorities must be set and promotion, tenure, and reward systems must be appropriate for the priorities of the system.

The challenge at the college at the University of Connecticut is the issue of teaching versus research as scholarly activity. How are scholarship and productivity defined? Definitions of faculty scholarship and productivity are essential prerequisites for instituting a meaningful recognition and reward system.

The college at the University of Connecticut is focusing on the definitions of scholarship as proposed by Boyer (1990):

- the scholarship of discovery;
- the scholarship of integration;
- the scholarship of application; and
- the scholarship of teaching or communication.

An adaptation at Oregon State University (see chapter 12) separates scholarship from research and teaching, stating that teaching, research, and extension are vital university activities that are not scholarship in themselves. Scholarship is defined as creative intellectual work that is validated by peers and communicated. This definition has been embraced by the College of Agriculture and Natural Resources at the University of Connecticut.

Such concepts provide for future directions. No longer is teaching pitted against research or research against extension. It is a takeoff point for rewarding faculty.

Cornell University (H. Dean Sutphin)

The College of Agriculture and Life Sciences at Cornell University is a large organization, with over three thousand students in twenty-two departments and sections. Like some of the larger colleges of agriculture in the United States, it has unique features. It includes the Division of Biological Sciences. It administers the major undergraduate business management program for the university.

At the college level, faculty committees provide a structure for policy development and faculty and student interaction. The college is linked nationally with the Kellogg Projects that include seminars to share ideas among a national group of faculty and administrators.

A review of the college academic programs was initiated in 1996. The first task was to define what the college should teach. The faculty defined and endorsed broad, conceptual outcomes of an undergraduate's education (see addendum to this chapter). These "educational gains" became the foundation for academic program review and development. Subsequently, the advising system was evaluated. The approach was to evaluate not how well the faculty was doing in advising, but how well students were entering into a relationship with faculty in the advising process. A committee of faculty and students defined the rights and responsibilities of the students in the advising process and the roles of the faculty: students have the right to an effective advising system, and the faculty should expect responsible action from the students.

Experiential learning was identified as an important part of undergraduate education. Criteria for experiential learning were established. The goal is to create hands-on, problem-based learning experiences to develop critical thinking skills connecting theory and practice. A strategic plan was developed for distance education. A strategic plan was developed for academic programs, including admissions, counseling and advising, career development, minority programs, student records, and registration. A set of principles was developed for budgeting allocation.

As the next step, tasks forces were created to review the academic programs. Representatives from across the university were included to build bridges to other programs. The focus is on four areas:

1. The teaching and learning environment to include task forces on students, library services, human diversity, and faculty development and evaluation.
2. The academic program support services to include admissions, financial and career development, student records, counseling/advising, and minority programs.
3. Interactive activities to include international studies, distance learning, electronic technology, and summer and intersession teaching.
4. Curriculum support to include task forces on course distribution requirements, core requirements, writing across the curriculum, ethics, computing competencies, quantitative literacy, cross-cultural experiences, and experiential learning.

Conclusion

Implementation of fundamental change in higher education in agriculture and natural resources is a shared responsibility of administration, faculty, and clientele. Administrative leadership appears to be key, but clearly faculty must have impact. Administrations should organize, seek, and use faculty thought and action.

The three cited examples are evidence of reaches for appropriate changes in their respective undergraduate and professional programs. The approaches taken by each tend to differ, dependent on the value system of the institution and the university environment of each college. Regardless of size of the college, the common task of colleges of agriculture is to manage multiple missions, but the mission and goals of the college must be described by the institution. The environment is increasingly requiring that more be done, but with less resources. The issues focus on balancing research with scholarship in teaching activities. An excellent course can be a valid scholarly effort, even though it may not be communicated to peers or published as a book.

Perhaps the most important need, relative to responsibilities and expectations, is to define relationships and institutional values within the multiple missions of the college of agriculture. The leadership for academic agriculture and natural resources must be given a high priority within the system.

It seems important that agricultural institutions be linked together to improve undergraduate courses as well as share concepts about the institutional support for undergraduate education. The three examples provided here emphasize institutional support. Just as certainly, we need to ensure that all learners see and understand the biological, ecological, and human sciences of the new agricultural and natural resource system.

Addendum: Cornell University College of Agriculture and Life Sciences Faculty Policy on Undergraduate Educational Gains

Rationale

Faculty, students, prospective employers, and the general public each have expectations of skills and traits to be gained from an undergraduate education. Most acknowledge the expectations that technical competence be acquired in major and minor fields of study. Increasingly, there are expectations that graduates possess a broad range of intellectual skills that allow them to synthesize knowledge, to achieve understanding of complex issues, and to contribute their specific expertise more effectively. The following policy reflects a consensus of faculty expectations of educational gains by all Cornell University College of Agriculture and Life Sciences undergraduate students. This policy is intended to guide the Curriculum Committee and others in the College of Agriculture and Life Sciences seeking to improve the quality of undergraduate and value of undergraduate education.

Policy

- The ability to understand and appreciate the complex biological, social, and physical interrelationships associated with the management of the earth's resources;
- the ability to write and speak effectively in the expression of disciplined thought;
- the ability to listen carefully and respectfully to the views of others, especially views with which we disagree;
- the ability to reason effectively in matters both quantitative and qualitative;

- the ability to access and make effective use of modern sources of information;
- the ability to evaluate and effectively interpret factual claims, theories, and assumptions in the natural sciences, the social sciences, and the humanities;
- the ability to communicate with people of different cultural perspectives;
- the ability to employ ethical reasoning in judging and acting on the moral implications of ideas and deeds;
- the ability to work both independently and in cooperation with others;
- the ability to evaluate priorities and to set and achieve goals;
- the ability to integrate theory with practice;
- the ability and interest to pursue lifelong learning.

8

Conception and Change

Key Concepts

1. The conceptions of the mission and purposes of higher education in agriculture and natural resources will drive change as much as any other trend in higher education.
2. Acceptance of their respective responsibilities by administrators, faculty and students is essential to implementing change. But, guiding conceptions are important in shaping the perceptions of such responsibilities.
3. Strategies for change will be based on defining new insights as well as the existing value system.

Introduction

The director of the project, H. O. Kunkel, dean emeritus, Agriculture and Life Sciences, Texas A&M University, in describing the processes in the project, suggested that the implementation of systemic change can occur if there are appropriate conceptual bases that facilitate change. In the minds of the participants in the workshops, the responsibilities of administrators to respect the teaching function, the importance of faculty to be concerned with institutional change, and the obligations of students for their own learning emerged as the largest factors in the revitalization of undergraduate education. These responsibilities are of primary importance in facilitating the makeovers that colleges of agriculture and natural resources need for the twenty-first century. Administrators, faculty, and students each have certain essential responsibilities in both the educational process and in opening the system to revitalization.

The Cardinal Step: The Conceptual Bases

There is an ultimate step, however, needed for fundamental change to proceed. The new directions that higher education in agriculture and natural resources can take rest in their educational program, that is, in the courses that the college itself offers and the curricular opportunities that the college constructs. It is the concept that drives these activities that must receive the continuing and ultimate considerations of faculties and administrations.

Therein lies the problem. Authorities differ widely as to what the major issues are. For example, some of the listed important concerns for agriculture that indicate the variety of viewpoints are

- the intensification of agricultural production, particularly in the less well-endowed lands of Africa, Asia, and South America;
- the threats of global climate change and population growth;
- the needed reorientation of higher education toward natural resource management and environmental quality;
- the rewards and risks of innovation; scientific data rather than intuition and emotion;
- preeminence of nutritional goals in setting agricultural research and education;
- the task of agriculture to provide more nutrients rather than more food;
- the environment as an irreplaceable resource; and
- change in education from a concentration on degree programs in the hope that the knowledge might be used later to education provided when and where it is needed, to customized educational services to meet the particular needs of students.

Curricula and courses in the agricultural and related sciences may also be fashioned in a specific perception of the purpose of agriculture as

- a way of life (the agrarian view and also a global view);
- an industrialized activity that is structurally in the economic base of the country (the industrial view);
- an activity visualized in societal terms as the means to human health and vitality and to environmental integrity and the risks in reaching these goals (the risk society view); or
- all of the above.

Each of these conceptions suggests a different set of course contents and a different configuration of the curriculum.

Additionally, the systems of agricultural production are dichotomous. Much of agriculture is technologically based, for example, crop, fruit, poultry, and swine productions. Others—ruminant animals, ocean fisheries, and forests—are largely based on natural systems in which the limits of production are now being reached. Of course, the technological and the scientific both border the natural systems. But human welfare and environmental integrity are imperatives in natural systems.

It is unlikely that the conceptions of education in agriculture and natural resources as well as the knowledge bases guiding course contents today will work in the twenty-first century.

Conclusion

The primary concept is the scope of colleges of agriculture and natural resources. We have defined the modern college of agriculture, food, and natural resources as

> the educational, scientific, and scholarly framework for the understanding, development, management, and use of biologically and ecologically based systems and relevant human resource systems for the benefit of human societies.

One of the cooperative colleges in these studies, the College of Agriculture and Life Sciences at Cornell University, has set the priority educational gain as

> the ability to understand and appreciate the complex biological, social, and physical interrelationships associated with the management of the earth's resources.

These statements are congruent, but they do not identify the environment of colleges of agriculture and life sciences in the twenty-first century or the strategies to facilitate change. Such factors will be dependent upon other defining insights.

The philosophical and practical consideration of the course and curriculum contents is critical. A guiding concept is that all higher education should prepare the student for work and life in what surely will be a rapidly changing society. Central to revitalization of higher education in agriculture and the natural resources is the connection of the redefinition of scholarship with the mission of the college of agriculture, life sciences, and natural resources. And again, as the community of higher education in agriculture and natural resources has not been standing still, some changes are being made. Such changes demonstrate the particular importance of the contribution of the individual faculty members to the insights of the institution.

9

Values

Key Concepts

1. Formal educational institutions—departments and colleges—have a determinate place or mission in the larger society. Educational institutions, including colleges of agriculture and natural resources, are structured, funded, and intended to produce college graduates or human capital for the vitality of a presumed successful socioeconomic system.
2. Although it may not be recognized, values education is already a fundamental part of science education because some level of professional, practical, or ethical values education occurs in every classroom.
3. Science has a product orientation that fosters the notion that the scientist must always be doing something. Public accountability dictates the measures applied to determine productivity of academic labor: publications, patents, productive uses, full-time equivalent student units (FTEs), and graduates. But the productionist orientation of scientists may create a conflict between academic research and teaching.
4. Among scientists, it is considered an ethical and professional obligation to give accurate information, to honestly report results, and to provide as many useful products and services as possible. Educating students and the general public about what scientists know is also a professional obligation. As the greater good is achieved through sharing such results, the agricultural science system has demonstrated a considerable degree of ethical legitimacy in utilitarian terms.
5. Despite their productionist tendencies, agricultural scientists know what processes can be productive. Communication of scientific knowledge is a valued and valuable process and activity. Moral talk can be even more productive if it is engaged. If scientists engage in moral talk among themselves, with students, with policy makers, and the general public, all involved may find it surprising how productive it may be.

Introduction

Institutional, scientific, and faculty values are critical factors in the implementation of change in higher education in agriculture and the natural resources. Values, in terms of this book, are beliefs about the desirable, in terms of which qualities, objects, and principles are considered to be of value. Attitudes, which also facilitate or inhibit change, are mental positions with regard to, or feelings and emotion toward, a fact or state. These combined with the ethics accepted by society form belief systems.

The issue is not the instructors' personal judgments. Surely, a faculty member with a distinctive belief system, such as vegetarianism, that of an organic food proponent, animal rights, environmentalism, an anticorporate position, or other belief, teaching that one set of conditions or another has a higher moral value, creates a mixture of personal bias, moral positions, and scientific values. But, individually or collectively, the presence of such moral positions on campus is not likely to enhance or inhibit the change process, although the positions are important issues for students to identify, understand, and critically analyze to develop a practical moral wisdom (Schillo, 1999). What is important is a belief system that accepts change in the curriculum.

Generally, modern colleges of agricultural and natural resources are partitioned along academic lines that no longer reflect intellectual life: modern knowledge systems are inseparably interdisciplinary. Sharing organizational insight and energy with the rest of the university community becomes a goal. This seems particularly true of the life sciences but now also includes economics, engineering, business, and moral philosophy. Among these disciplines, moral philosophy remains novel and variably included in the academic scope of higher education in agriculture and natural resources.

The considerations of values in higher education in agriculture and natural resources, are twofold.

1. Values permeate the scientific community, and hence values permeate the teaching of science including the agricultural sciences and natural resources.
2. But among the values that permeate the scientific enterprise, there is a profound distrust—or at least skepticism—about the appropriateness of ethical or moral considerations inside science and scientific education.

If the latter point can be somehow turned around in higher education, then it may well create the openness that will allow examination of the values that direct our courses and curricula and thus facilitate implementation of change in higher education in agriculture and natural resources. This possibility was examined in detail by Jeffrey Burkhardt, director of the Ethics and Policy Program, Institute of Food and Agricultural Sciences, University of Florida, and through considerable discussion among participants of the Texas workshop.

The Nature of Education

Burkhardt first noted that education is not solely about the transmission of information, knowledge, and skills. It is at least as much about shaping, transforming, enriching, and disciplining people. Burkhardt alleges that too often teachers in colleges and universities forget that regardless of whatever else they do, they try to change people by changing or disciplining their minds. The frequent lament of teachers not "getting through" to their students only reinforces the idea that values of various kinds permeate the learning environment.

Values are inherently part of the instruction provided by faculty in the agricultural sciences and natural resources. However, the values instruction, or transmission, which occurs in agricultural and natural resources is usually not acknowledged by faculty members to be values instruction. The fact that values are not acknowledged to be values can lead students to believe they are fixed "truths."

Education, however, takes place in institutions. Institutions such as family, town, or church are fundamentally instrumental in transforming people, but for the large majority of young people, formal educational institutions—schools and colleges—have a very significant role in determining what people know and who they are. The whole enterprise is and has been deemed socially useful.

Burkhardt argues these points: Education is about changing people, and thus the process occurs most often in a context. This point is clear even if not constantly acknowledged. As a result, values of a variety of kinds are intrinsically part of the educational system and process.

Values in Science

Higher education in agriculture and the natural resources is rooted in science, particularly, the biological and social sciences. But the value system in science in higher education is not always clear to either students or graduates from the system. In the overidealized, undergraduate textbook, homogenized definition of the scientific method, it would seem that "values" have no place in science. Burkhardt notes that if all there is to science or being a scientist is rigorous and impersonal application of the scientific method to whatever phenomena the individual is intent on studying, then it would seem incredulous to say that values enter into the process. This thought is carried on in some arguments relative to the debates about regulatory matters, where scientists argue that regulatory issues must be settled by science. The argument is that conjecture, or "emotion," should have little place in regulatory decision. Burkhardt notes that under such conception of "good science," personal and cultural biases or unfounded opinions are or should be out of place. However, Burkhardt also suggests that there are other commonly shared values within the scientific community.

If science is also about a complex set of actions, judgments, behaviors, disciplines, and contexts, values are clearly a part of science. Science is people doing a myriad of things. So long as people are doing, thinking, and saying things in this enterprise, values are fundamentally there, if for nothing else, as Burkhardt concludes, than people have values and act upon them.

Values in the Classroom

Burkhardt continues that certainly, in higher education, there is teaching and learning of theories, facts, formulas, procedures, and the other routine or mundane activities. Consider the underlying message in these ordinary activities. That they are taught by a professor, a professional scientist in many instances, reinforces for the student that these learning activities "really count" as being scientific or a part of science per se. A subtle but persuasive activity is occurring.

In the process of doing and teaching the sciences, people are teaching and learning ways of seeing and ways of being. Teachers impart—and students take in—lessons of technical appropriateness. But they also internalize other perceptual and attitudinal norms. Deep learning and deep reinforcement of beliefs and attitudes occur along with the transmissions of technical facts and methods. Therefore, values education is already a fundamental part of the science education enterprises; practical or ethical values education occurs in every classroom. The question is, What values are being transmitted to students?

Value Education for Agriculture and Natural Resources

There is a lament among older scientists and administrators in the agricultural sciences regarding education: the concern is that for a variety of reasons, students in the agricultural sciences do not learn about the values of the land grant mission, that is, the fundamental social importance of agriculture and the role of the land grant university and its research and education to support the food, agricultural, and natural resource industries. Failure to reinforce these values and the land grant tradition among faculty and students may result in higher education in agriculture and natural resources losing its social and historical significance and justification.

However, taking a key thought from the first phase of this study (Kunkel et al., 1996; part 1), Burkhardt suggests that the concern may be akin to lamenting the passing of the horse-drawn buggy or the traditional family farm. He lists three factors:

1. In most colleges of agriculture and natural resources, the percentage of students either from farming families or who intend to go into farming after graduation has rapidly declined. Career interests are directed to professional careers or industrial agriculture.
2. Younger faculty members in agricultural colleges are increasingly neither from agricultural backgrounds nor educated at land grant universities. They are employed to engage in science. They want to teach science. They have little history or interest in the land grant tradition.
3. Consideration of the social value of agriculture or the land grant service mission does not generally fit the agricultural sciences because consideration of these values leads to questions of ethics or the social responsibility of science and scientific institutions. That would suggest to many faculty members that moral and ethical concerns are fundamentally precluded from the agricultural science classrooms.

Colleges of agriculture and natural resources have thus institutionalized science and science education. This has placed two belief sets into the system, both of which are long identified and long criticized: "scientific productionism" and a positivistic theory of knowledge. Scientific productionism is the idea that science should continually produce more output. Positivistic theory includes the idea that science is value neutral and the idea that knowledge is attainable only in certain ways and only with respect to "facts." Even if it is not true that all science and science education is productionist and positivistic, enough of it is to influence the scientific enterprise.

Science and science education are essentially result-driven enterprises. Scientists have long been seen as performing an important social function: to generate knowledge to improve the well-being of people, with a particular eye toward increased human control over nature and natural processes; that is, the production of scientific results has a distinctively moral purpose. Scientific activity in the agricultural and other sciences as well as scientific communication should be useful, even if only to other scientists. Basic science also fits this definition because the knowledge advanced by basic or theoretical science holds potential real-world use value.

Burkhardt does argue that there are subtle and mundane ways by which "scientific productionism" operates, which may undercut the long-accepted moral justification of agricultural science. Productionism in agricultural science, usually interpreted as a goal of greater yield and economic efficiency, may be tending to undermine its own justification by eliminating its "greatest social good" as well as the variety of scientific and nonscientific means of achieving it.

Quantity of Knowledge

One of the results of productionism in science is that the pace of scientific output has intensified. Professor LeRoy Klein, Biochemistry and Macromolecular Science, College of Medicine, Case Western Reserve University, joined other plenary presenters in the workshop at Cornell University. His discussion contributed to considerations relative both to values and science and to the problems of evaluating scholarship. Klein's focus was on the "avalanche of information" and the need to integrate that with both old and new problems. Although Klein's interests are directed to the biomedical sciences, the relevance of his comments to agricultural and natural resource science was evident. Agriculture, natural resources, and biomedical sciences are derived from fundamental biological principles. But the nature of the biological sciences is playing a role in determining the kinds of faculty in each kind of college, be it medical, agricultural, or natural resources.

The "explosion" of information is creating fragmentation in the understandings of science. Scientific groups may live in their own scientific compartments (Klein calls them "foxholes"). Groups in one foxhole are unaware of what is going on in the "foxhole next door." Now we have the Internet, which provides millions of pages of information, but the Internet provides no solid means to integrate and then to evaluate the good papers or bad papers or which are the "soft" data and which are "hard" data. Most data on the Internet are soft, but science is interested in the hard data.

There is an additional problem—the proliferation of scientific papers, that is, the productionist character of science described by Burkhardt. Klein, however, asserts that the problem of fragmentation is exacerbated by current facts of funding, and with fragmentation, our goals and principles become diluted.

Klein noted that growth in information as a field is linked to funding patterns. Nationally, there have been relatively small increases in funding and, hence, in the accumulation and flow of information in chemistry and physics; so is it also with business and sociology. Agricultural sciences have had some increase in funding, but it is more complicated as the increases are in applied areas, such as human nutrition, conservation, and pollution. There have been logarithmic increases in biological sciences and medicine due in part to the great amount of information in biology that remains to be discovered. Even so, some areas of biology are being lost: anatomy, physiology, and animal nutrition. Faculty members in these basic sciences are learning that even though they are good teachers, they can no longer get sufficient funding for research. With continuous restructuring, faculties may be developing that are heavily skewed toward molecular science. This, in itself, has value because the biotechnologies offer new interventions in agricultural processes, but molecular biologists do not usually integrate knowledge at higher biological and practical levels. A void in academic working knowledge is being created.

Burkhardt notes that the public views the production of human capital as the most striking aspect of productionism in science. But without integration, the sheer volume of information is the source of increasing complaint from students and teachers.

There is the thought that undergraduate professional programs should move from four to five years. Managing students—advising—has become a more complex task. Administrators in many colleges and departments worry about how to maintain quality in their educational programs given the quantity of knowledge that must be communicated on the one hand and the relative scarcity of scientists available (or willing) to communicate that knowledge on the other.

Beyond Scientific Knowledge

In science, rarely are questions of larger social and scientific responsibility fathomed in a systematic way. A major exception is the series on science and society that appeared in the journal *Science* through the years 1998 and 1999, but at a working level, talk of morality has long been considered inappropriate in the scientific enterprise and especially in science education. Burkhardt's argument is that this educational position is self-reenforcing when combined with productionist ideology. Moral talk is considered by some to be inappropriate because there are "no answers," only "opinions." It would follow that, if there are no answers, only opinions, then moral discourse is fundamentally unproductive. A problem is that some so-called moral talk is, in fact, only opinion.

Burkhardt observes that among most scientific professionals, it is considered an ethical obligation to give accurate information, to honestly report results, and to provide as useful products and services as possible. Educating students about what professionals know is also a moral obligation. As the greater good is achieved through the process of widely sharing the results of research, the science system has demonstrated a considerable degree of ethical legitimacy in utilitarian terms.

However, graduates are produced by the science establishment with little knowledge, appreciation, and understanding of the moral quandaries that they may encounter in their professional and business careers. With little or no knowledge of the utilitarian morality of science, of the moral issues that they may face in their careers, graduates may be left to rely on their own uninformed devices or opinions. This is hardly consistent with the utilitarian ethic behind public service. Just as the production of inferior products can give cause for questioning the legitimacy of the scientific results, the production of inferior products of education should also be questioned.

This is not to say we should abandon scientific productionism. Promotions, salary increases, and other academic incentives are tied to the production of science and product. But there can be time for thinking and talking about the ethics, values, context, and consequences of science. This requires recognition of various uses of morality as a productive enterprise. Abandoning scientific productionism is not suggested. It is expanding the goals over which productionism operates.

Conclusion

The scientific enterprise is not only about discovery of the world out there. It is also about the hearts and minds of those of us who participated in the workshops on the implementation of strategies for systemic change in higher education in agriculture and natural resources. Agricultural sciences have a moral purpose. Part of the moral purpose is the process of living with others both inside and outside science, utilizing the best and most sophisticated tool we have at our disposal—our common moral knowledge. The agricultural sciences and natural resources have another moral purpose: the greater good of society. But the sciences have been dominated by two tenets: (1) the idea that science should continually create more output and (2) the idea that science is value neutral. These forces have led to the long-held idea that talk of morality is inappropriate in the scientific enterprise and especially in communication of science and education. Although we may not have complete moral knowledge, codified in textbooks or Web sites or available on CD-ROMs, there is a process of sharing what moral knowledge each of us has acquired. This can bring a deeper qualitative dimension to our students' productive lives. Their sharing of moral "theories," data, and bona fide knowledge can enrich our own productive lives.

Students might take a course in ethics (agricultural and environmental), or faculty members may seek out moral philosophers as consultants. But the greatest change will likely come when the faculty members recognize, examine, and talk about the fact that morality drives what they teach. Issues of moral philosophy should become a greater part of the academic effort. Out of this can come the visionary purpose of higher education in agriculture for the twenty-first century.

There is also a moral aspect in the domination of research funding, which is perceptually structuring a faculty that may be losing the fundamental skills, desire, and time to integrate knowledge. Funding available to scientists will affect their productivity. If a faculty member is employed with the purpose to access research funding, then something vital to everybody's future may not be attained.

10

Course Construction

Key Concepts

1. The curriculum and, more so, the courses, their contents, and the structure of the faculty that teaches them are the framework through which change can be implemented.
2. The course should be appropriate for the problem or situation from which it derived. One cannot know if the course is appropriate unless one knows the problem or situation that generates the need. Courses should be outcome, not content, driven.
3. The course should have a clear and comprising rationale. The rationale should provide the argument as to why the course goals are important and why they are specific. Specific course goals are important because they provide students with the competencies they truly need.
4. The course should teach content that is necessary to accomplish course goals.
5. The course should give the student opportunities to practice what the designer of the course intends the students to learn to do. The course should provide opportunity for students to actually learn to analyze situations, to solve problems, to design and make decisions, and to evaluate products. The urge to "cover the subject" can be a negative factor in providing the practice the student needs.
6. The course should evaluate students in ways that are consistent with the intent of learning outcomes and provide timely feedback to students about their progress. Authentic assessment challenges course designers to produce evaluations that are more like reality and less like proxies for the real world.

Introduction

George Posner, professor of education at Cornell University, provided and led the discussion from which this chapter is mainly drawn.

It is through redesign of courses and curricula that fundamental change is implemented. The curriculum provides the framework. But even when significant change is visualized, the existing curriculum will likely be repaired, remodeled, or added to but not totally reconstructed. Curricular and course changes occur with the definition of a problem or issue. Posner suggests that the description of the situation that needs to be addressed will likely vary with whomever is asked.

The faculty will likely define the problem that needs to be addressed in course redesign when the curriculum is out of date, standards need to be increased, or there are new areas of knowledge that are not yet taught. Such definition is a highly content-driven formulation of the problem.

If employers and professional and graduate schools are asked about the adequacy of the courses and curriculum, often times the response is that the problem is an inadequate preparation for work or that students are ill equipped for further professional development. This is a more outcome-driven formulation of the problem.

If students are asked what they think about the adequacy of a course the responses many times are that the students cannot make connections. The curriculum is too compartmentalized or is not practical enough. The tests do not measure what they learned. The problems for students are experience, or process, driven.

Ask administrators and the response relates to cost or duplication of courses. The issue for administrators is resource driven, that is, the responsibility for resources and their distribution.

If the problem is formulated as content driven, the solutions appear as increasing requirements, adding new courses, and updating the curriculum. If the belief is in outcomes, solutions will be such as needs assessment and employer surveys. If the problem visualized is more process, or experience, driven, interdisciplinary courses, experimental courses, and better evaluation techniques may appear to be solutions. If the problem is wasteful duplication, do a resource analysis. The formulation of the problem leads to the design of the solution.

Curricular Design

Implementation of curricular change can be viewed in two ways.

1. The curriculum is seen as a means to disseminate new knowledge. It is the supply of new knowledge that activates the process of change.
2. In contrast, the curriculum may be seen as an instrument for goals and meeting needs.

Posner argues that it is the latter purpose that should guide curriculum development in colleges of agriculture, life sciences, and natural resources. The instrumental purpose of accomplishing goals and meeting needs broadens the definition of the subject matter. For example, students in animal science need more understanding about soils and hydrology to increase emphasis on environmental implications of animal production. Now, what was previously more narrowly defined as animal science is broadened because of the needs that are out there.

Curricular design may reflect macrolevel planning, that is, college- or universitywide planning. It may reflect such concerns as the writing capacity of the students or their mathematical competencies and the level of the course. Microplanning is where curriculum development is targeted at a particular unit in the college. This may involve a particular major, design of introductory courses, design of capstone courses, concerns about sequencing, preregistration requirements, and so on. It is in such a framework that thought may be given to the design of a course.

Curricula are the instructional plans that are the means of implementing intended learning outcomes. They are the means of implementing goals and telling us how to accomplish them. Curricular planning is not a linear process. In planning a curriculum or course, thinking must move backward to the justification and then move forward to the structural plan.

Course Design

The elements of course design, as Posner focused on them, guide also the conceptual bases. Educational goals are formulated and expressed in the context of the rationale of the course. The rationale expresses the questions, Why these goals? and Who is the course for? The rationale expresses the intended audience and the values that are implicit in the course. Educational goals can be the expected educational gains but need to be stated in course-specific ways. They are the competencies that are expected to be accomplished in the course.

The intent of the learning outcomes is the "meat" of the course design. This can be expressed as objectives or in the actual course content. Posner suggests that a good way of graphically organizing the intent of a course is a concept map. Concept maps can show the coherence of the content, and they can be fairly simply constructed. The concept map also helps to ensure greater continuity within the course and to expose possible gaps.

Another way to develop coherence in the course is to develop key questions, but no more than three or four. If as an instructor one gets a better sense of the whole, the students will also get a better sense of the whole. Key questions are a good way of finding ways for a class to reach coherence.

Course design also includes instructional planning. An important principle in instructional planning is to focus less on what the instructor is going to do and more on what the students are going to do. Students learn principally by what they do, not what the instructor does. If a student is to learn to analyze situations, solve problems, evaluate products, and think critically whatever the goals should be, students need the opportunities to do these things. They need to manipulate information in certain ways. If teaching is by generally straight lecture, the instructor has to ask of himself or herself what the students are doing while the instructor lectures. Probably, the skill of the instructor determines what the students are doing, but the question itself can be telling.

Posner advises that instruction should be planned on what we believe students already know. No student is an "empty vessel." The original project work group identified the danger that contributions of stated prerequisites to a course are often ignored by both faculty and students. A written exercise at the beginning of the course is suggested by Posner, not to find out the ways that students are wrong, but to find out what they think.

The last element in Posner's thoughts on course design is evaluation of the students. Student evaluation aligned with course purposes can be a large boost to the effectiveness of the curriculum. The evaluation plan should indicate the ways in which both the students and the instructor will receive information about the progress that they are both making. Evaluation of students should be timely and accurate, but it does not need to be all in the written test; there are other ways of obtaining feedback. The most important point, however, is that the evaluation technique should be aligned with what is intended. It is the indicator of what students focus on. If the instruction is aimed at higher order thinking, reasoning, and analyzing situations but the evaluation techniques are aimed at recalling facts, students will shortly determine where to put their efforts.

The Framework

With the rapid accumulation of knowledge, the temptation for the instructor is to resort to "covering" substantial content in the course. Posner argues that when one tries to cover tremendous bodies of information, then there is only one thing the instructor is going to do: cover the content. Students have no way that they can "dig" into the content and really think about it if every time they turn around there is more. Thus, teachers must set constraints on themselves. Some serious work needs to be done to determine the key concepts and how the other parts of the content are derived from them.

There is always a trade-off between depth and breadth in a curriculum because there is a limitation of time. A decision has to be made. The movement in the future appears to be more and more in the direction of breadth rather than depth.

Conclusion

The basic framework of good educational design is the same regardless of the level—a particular course, a major, a department, or a college. The thinking should be on educational goals, not in the specific, but in the larger sense. Thinking still has to be about what it is that the students should be learning: major ideas, the major issues with which one should be comfortable, and the skills one wants them to learn. In each case, we are going in and talking about, at least at a general level, the approach to instruction.

In course design and in presentation, reflective thought is necessary. Such thought would include the way students are to be evaluated. During the course, comparisons will be made between the educational goals and the actual educational gains.

This is the framework in which a course should be designed: a shift from inputs (faculty) to outcome (students) should open the door to systemic change in higher education in agriculture and the natural resources. Preparation of the course is key in educational reform.

However, there is a current in many institutions: that is, overloading courses, both in number of students and in content. As new knowledge develops, such as it is with molecular biology or economic theory, the tendency is to add courses rather than revamp the current ones. Undergraduate students will give only so much time and effort to their undergraduate careers. However, relevancy, systems thinking, and other such aspects of modern higher education in agriculture and natural resources can be integrated into current courses, and additional courses may not be needed.

11

Course Content

Key Concepts

1. The contents of courses and curricula in agriculture and natural resources in past decades were generally reflective of the society that students were expected to enter. But significant changes have occurred and are occurring in North American society without a concomitant level of change in programs and course contents of higher education in agriculture and natural resources. These kinds of academic changes are a major component of the needed fundamental change in higher education in agriculture. The greater challenge is to provide a new mental model for education for the twenty-first century.
2. Higher education in agriculture and the natural resources today is largely based on the perception of work and life in the industrial society. This reflects the notion that we educate students to obtain a given status. Students are given the technical skills to be able to control their own benefit streams: employment, income, health benefits, and quality of life. Life-time stability has been expected.
3. But stability cannot be counted on. Society has changed considerably since it settled into an industrial organization. Political and economic issues reflect a dependence upon fluctuating alliances among interests. The current preoccupation is with risk.
4. The first requirement of education for a risk society is to convince students of agriculture and natural resources that a risk society with global dimensions is here.
5. The key to the higher education in agriculture needed for the future is this: higher education should prepare students with the technical and communication skills plus background knowledge that they will need for initial employment *and* to be able to handle the problems of society, develop values, and apply these values with respect to the multiplicity of risk issues.

Introduction

The controlling issues in revolutionizing higher education in agriculture and natural resources in the future relate to the control of the curricula of its undergraduate and professional students. Curricular contents are often controlled with the corporate affirmation of the faculty. Course contents, however, are usually the prerogative of the individual faculty member. The exceptions are underclass (freshman and sophomore) courses when tradition and available textbooks dictate the content.

Professor Paul Thompson, Department of Philosophy, Purdue University, has been a key member of this team concerned with systemic change in higher education in agriculture since the inception of the project. He has a firm sense of the tracks that the thinking has taken. He was invited to prepare the presentation on course content, not as a summary of thoughts during the years nor as a discussion of what he has done in his courses, but to provoke the broader discussion of the key issue, the intellectual control of course contents for the future.

Thompson's stimulus is derived from the book by Ulrich Beck (1992). Thompson states that as a result of listening to others who have written about challenges to agriculture, he believes that there can be new ways to think about the challenges that our graduates are going to face. Thompson framed his presentation in terms of the notion that society is in transition beyond its industrial organization. When we think about preparing students for careers or jobs, whether it is to be a producer or an employee in an agricultural supplier or in a marketing system, we have a kind of mental model of how society works. Thompson suggests that not only is that model not really working anymore but that young people already have a different mental perception for the relationship between their education and their working environment after graduation. But many faculty members in higher education in agriculture face challenges because their own mental perception is a model fixed in the industrial mode. There is concentration on production, emphasis on efficient uses of resources and of labor, and emphasis on greater efficiencies in terms of the wave of different technologies and different sectors being integrated together. Different priorities may exist in the future.

Transitions: The Trajectory into the Future

Characteristics of each society—agrarian, industrial, at risk—are relevant to how students are prepared in higher education in agriculture. In an agrarian society, a person's life prospects depend on relationships to others who are well known to the person. They are family, close neighbors, and people in church and some other local organizations. Life prospects are tied to people with whom one has face-to-face contact. What you are and who you are matter very much. All of these are context relationships and if one goes away to a university, he or she will be getting generalized, abstract sorts of relationships that are likely useful only to the few elite.

For agriculture, a transition to an industrial society occurred during the years leading up to World War II and immediately thereafter. What distinguished the industrial society from an agrarian society is that one's life prospects depended upon one's ability to control benefit streams. The benefit streams include more than income because the benefits from the job are broader than income. The idea is that there is a general class of benefits that flows in a continuing fashion, and benefits come to an individual as a function of his or her status in society, his or her position, job, or whatever. Status would also respect specific confidences one has as an individual. Thompson states that by status, we think about needing an employee who occupies an investment position. By confidence, we think training in job skills and knowledge. Higher education's job became one to make sure our graduates come out with the competencies that would allow them to eventually occupy one of these status positions and give them control of the benefit stream. Some of the benefit streams are wages, profits, retirement funds, capital gains and entitlements including subsidy programs, welfare benefits, and education.

The industrial society is a society that is primarily defined as competition. Individuals are primarily engaged in competition for controlling the benefit streams. This tends to land strongly on class lines. One can see the interests of labor and the interests of management and the people who hold investment as representing different classes. A look at the history of the first half of the twentieth century reveals a lot of social movement and of government policy being dictated by class conflict. It was less true of agriculture because agriculture resisted the transition to an industrial organization until after World War II.

There was a stability in the industrial society. The presumption in the 1940s and beyond was that the family home would be stable. Marriages last a lifetime. People live in the same home all of their lives or make only one or two moves. Surrounding neighbors are still there, but one does not need to rely on them in order to succeed. The life-long occupational status prevails.

Gender, race, religion, and national origin were important in terms of fixing a person's status, and as it became difficult in the industrial society to move across some of these boundaries, they became ways that the classes defined themselves.

In the industrial society, the assumption is that public health would gradually improve, through both medical discoveries and an increase in the benefit streams in terms of public health. With regard to nature, there is a general assumption that nature is invulnerable and we can do little to hurt nature. Nature is so much larger, so much more comprehensive than human activity, that it is not really in the thinking mentality of the industrial model. And there is also an assumption that "just as I have done this all my life, my children will be able to do it all their lives."

Thompson suggests there are also some important elements of change that continually take place within an industrial society. There is class competition for controlling the benefit stream. This competition generates changes in public policy and changes in the work process that tend to destabilize relationships and force individuals to be prepared for some of the change. In this instability, there is a kind of stability. The issues in ideology all came under a general framework of how far government should go in terms of moderating or trying to affect the competition over income streams. There has always been one part that tended to think the role of government is essentially interfering in this competition, and there has always been another party. Defined by competition, it is clear what is on the left and what is on the right.

This is the educational view existent at the beginning of the twenty-first century. Change occurs also in terms of technology. New technologies have always come on-line. In many industries, including agriculture, there will be changes in production methods and new products coming along. The new product may not be a direct replacement of the existing product, but it may displace it. Awareness of the technological changes and knowledge about technological changes are a crucial part of the life skills, particularly for people who are in executive positions, including the agriculturist who must decide what technology to use in producing crops or animals. Although technological change has been a major part of the mental model of living in an industrial society, there has been a generally receptive attitude toward technological changes even though they did introduce a certain amount of instability and created a need for educational responses within the industrial society model.

The Risk Society

Thompson, citing Beck (1992), suggests that the terms liberal and conservative make little sense now because both the liberal and the conservative win or lose as a result of technological change. Class does not fade—workers as a class, farmers as a class, women as a class, Blacks or Hispanics as a class. But what technological changes do is affect the individual differentially. Conducting politics in terms of large class movements only works in the form of temporary alliances that focus on a specific policy or a specific initiative. People mobilized in a group do not share many interests other than one issue, such as improving children's health. As soon as the focus is on another issue, the alliance breaks up.

One area of stability in the industrial society was the home and family. In today's society, people have multiple homes and the trend is toward multiple successive partners. It seems that pressures are just part of living as society intends to make all sorts of alliances temporary as opposed to life long.

The idea of a generally improving public health was accepted in the industrial society. Public health does seem to be improving in the risk society, but that is not the perception of most people. People see environmental and dietary risks as a significant reversal of any trend of generally improving public health. Thus, the assumption that technology will lead to a general improvement of public health is no longer operative.

As for life-long jobs, students should not expect to spend their lives in the same career, much less in the same job. They can expect to work for two or three firms during their first ten years after college, and then, probably having played out a particular career life, they will have to find something else to do. One hopes that they build on some of the skills they have developed in their first years after graduation. Students should see their future life as a movement from one job to another, and they can see the positive aspect of that. It goes along with a stable job hierarchy idea: certain kinds of people tend to occupy certain jobs, and once they get into that status, they are largely set.

Turning toward nature, the perceptions are opposite in the industrial and risk societies. Whereas we once thought that nature was invulnerable, we now think of nature as being vulnerable.

In the industrial society, politics was a stable contest between left and right and was dominated by large groupings of people into a class of status occupations. Risk politics is described by Thompson as largely an issue-to-issue competition over who will bear the risk and who will be the one who does the thing that poses risk in a particular situation. In this atmosphere, people will work together in one situation and against each other on another issue.

In agriculture and natural resources, there are expectations of what people are going to do to challenge the whole model. There are issues: genetically modified organisms, growth factors, pesticides, industrialization of agriculture, organic agriculture, food safety, and others that tend to create those clusters of parties whose interests are not economic. It is not an employee benefit stream interest, but a specific risk issue, and those groups will fluctuate and change over time. Certainly with an environmental risk associated with technology, one cannot define winners and losers on a class basis. People in all classes both benefit and lose because of environmental manipulation.

Course Content for the Risk Society

Some of the issues the risk society draws attention to in higher education in agriculture and natural resources are food safety, human nutrition, air and water pollution, resource depletion, land use, climate change, and occupational health and safety. We argue that these are all issues that are associated with what an individual does and what society does with respect to each of the issues. These are not general class issues, but issues about who is going to do the risk and who is going to bear the risk, about compensation, if any, or litigation, or about possibilities of working together to evaluate the risk. There are issues of animal welfare, vegetarianism, and dietary rights (sometimes related to animal welfare.) These are food scares (mad cow disease, genetically modified organisms), race and gender discrimination issues, challenges to property rights, and issues of quality of life in rural areas. But just because these issues are on the table does not mean the old issues of "How am I going to stay in business?" are off the table. The old issues have not gone away. One cannot think educating people for the risk society somehow means abandoning the old curriculum that produces technical confidence. Those needs are still there.

The added point is that it is not enough for one to sit on his or her own intellectual resources. The broader public must be engaged on all of these issues. One of the competencies that higher education has to produce is the ability to communicate, which means the ability both to speak to and listen to the public with respect to these issues.

Society is in transition, and possibly the best definition of it is that it is a risk society. Risks exist in both developed and developing countries. The first task of education for the risk society is to convince agriculture and natural resource students that the risk society is here. Many students are ahead of the faculty in this regard, but Thompson senses that many students only want the technical competencies that are needed to manage an operation for some company. They still think of the industrial model centered on life-long jobs. Actually, some students and faculty still cling to the agrarian model.

Part of the education in the face of the transitions that seem to be continuing—part of agriculture as a way of life, part still functioning under the industrial model, and part in transition from the industrial to the risk society—is to prepare students with technical and communication skills plus the background knowledge that they will need to be able to cope with and handle the problems of society and develop values with respect to the multiplicity of the risk issues.

Most faculty members educated after World War II were educated in a system preparing them for the industrial society: preparing to be a producer and a farmer for a long time, preparing to be an employee and then perhaps moving up to executive status, or perhaps preparing to be a college professor. Faculty members acquired the knowledge and technical skills that they needed to be able to get into that position and stay there. Some faculty members are still struggling with the transition to the industrial model as opposed to an interdependent family farm model. It may not be easy to convince faculty that the risk society is here.

Thompson returned to his original point. Course content remains primarily an individual decision of the faculty member. This is based on the notion that academic freedom must prevail. If the faculty cannot be convinced of the needs stated here, then we have really no access to the problem of course content.

Conclusion

Instructors in higher education in agriculture need to take a close look at what they are doing in the classroom. Many courses in the agriculture curriculum are clearly pointed toward a graduate successfully carrying out a career under an industrial model. The people who teach such courses need to evaluate what they are doing and whether or not they are truly preparing students for the world in which they will live. That does not mean teaching ethics or sociology or something that is not part of their subject matter, but they should rephrase the subject matter, even if based on natural sciences, to address the rationality of a rapidly changing world of work. For example, in an animal nutrition course, say as much about human diets as about animal diets. The essayist (Paul Thompson) argues that students ought to be coming out of college today with the sense of how they are moving toward the risk society and how that differs from education for the industrial society model. The current transition in the society will take us beyond the industrial society, but it is likely to be a long transition, throughout the lifetime of everyone who is graduating now.

Four points seem to be imperative to the implementation of reform in higher education in agriculture and natural resources:

1. The rapidly changing nature of society of the twenty-first century must be understood. Its nature must be defined.
2. Students must be convinced that a rapidly changing society exists.
3. Graduates should be prepared for a global economy.
4. Faculty members must be convinced that a changing and global society exists, and the university should be preparing them as well.

The implications of differences between twentieth and twenty-first century food systems, that is, the industrial agriculture and agriculture as a risk society, if that best describes the twenty-first century society, and recognition that a graduate will likely have multiple careers may be lost on students as they come to our universities. Similarly, students may not be fully aware of the implications that these factors might have on the course work they are taking and the knowledge they should be acquiring.

We can put these issues in twenty-first century terms. Corporations are employing graduates of colleges of agriculture for a much wider range of positions and in increasing numbers. Information technology, enterprise technology specialties (agricultural economics, agronomy, animal agriculture, entomology, forestry, nutrition, biotechnology, operational engineering, agricultural business, etc.), and management of integrated units are providing the greatest number of opportunities. Presentation skills (written and verbal), ability to think logically, analytical skills, sound judgment, ability to work in multicultural teams, and a driving thirst for learning are skills that ought to be goals of higher education. At center will be the ability of graduates to change with the organization. The student needs to cultivate an ability to learn new skills, be adaptable, be curious, and have a love for learning.

The role of the faculty is to ensure that students are being equipped for the twenty-first century society, not for today, which might involve roles for which both faculty and students have not been specifically educated. The foundation may be laid by learning relevant fundamentals, experiential learning, case studies, and studies of contemporary issues.

12

Scholarship

Key Concepts

1. Scholarship is required of essentially all university professors who are on tenure track. Traditionally, only research has been recognized as scholarship in evaluating faculty performance.
2. Change in higher education will most likely occur when a broader vision of scholarship recognizes scholarly contributions in the teaching, learning, and the pedagogical processes.
3. When all forms of scholarship are recognized, including the discovery, integration, dissemination, and application of knowledge in evaluating their performances, faculty members can be appropriately rewarded for their creative scholarship as teachers.

Introduction

The general, current system of rewards and incentives for faculty members are based on a value system that is common to essentially all North American universities and, hence, to colleges of agriculture, life sciences, and natural resources. The value system derives from the values of the science enterprise.

Jeffrey Burkhardt's (see chapter 9) discussion of values is pertinent to this discussion. Burkhardt, C. J. Weiser (Oregon State University), and other participants observed the well-known truth that scholarship in academia is focused on research. Burkhardt observed additionally that sciences have a product orientation. The obvious kinds of science output are new technology and patentable products and processes, published materials, and other physical products of scientific work. Less obvious are administrative reports, grant proposals, workshops, media statements, congressional testimony, extension training, and consulting. Less obvious still—as products—are the postdoctoral trainees and laboratory assistants and graduate and undergraduate students educated at our universities.

It is the more obvious output that is the major factor in recruiting and rewarding faculty members. The valuations of physical products of science, particularly publications, often outweigh the evaluations of teaching and outreach, probably because they are more easily documented. Research publications in refereed journals are validated and evaluated by peers. Peer evaluation of teaching and outreach has been less formal in universities. Teaching portfolios and peer reviews are becoming widely used to evaluate teaching. Student evaluations are often factored into the evaluation process.

Scholarship remains a common, if unstated, component of the teaching process. Scholarship has, however, often been narrowly defined as research or discovery of new knowledge in most reward and incentive systems for faculty. The discovery associated with research has been the most acceptable component. The late Ernest Boyer (1990) said that forms of scholarship were discovery, application, integration of knowledge, and teaching. Scholarship was undefined but described as things that professors do. Hence, good teaching could be considered scholarship.

In *Scholarship Assessed* (Glassick et al., 1997), published after Boyer's death, it is suggested that scholarly achievements can be assessed by determining that "phases of an intellectual process" occurred during their development. The six sequential phases of an intellectual process are described as: (1) clear goals, (2) adequate preparation, (3) appropriate methods, (4) significant results, (5) effective presentation, and (6) reflective critique. Boyer's great contribution is the concept that scholarship goes beyond research and includes other types of intellectual work.

The Oregon State University Model

In 1995, Oregon State University (OSU) became the first major research university to utilize an extended concept of scholarship in its revising tenure and promotion processes. Kent State University, Portland State University, Montclair State University, the University of Idaho, and Iowa State University have made similar changes, and discussions are underway at dozens of universities. Conrad J. Weiser, dean emeritus, Agricultural Sciences, Oregon State University, provided discussions of the broader vision of scholarship, which provided the conceptual foundation for revising promotion and tenure guidelines at Oregon State University.

Weiser began his discussion with four assertions:

1. Parents of our students, employers of our students, and taxpayers that pay our salaries, for the most part, disagree with the value system of universities. The perception is that universities are driven by research grants and research funding and have little interest in undergraduate education.

2. The clearest indicators of what the value system is at a specific university is not a statement of strategic positioning, its mission statement, or its strategic goals, but, in fact, it is the criteria that the university uses to evaluate its faculty, specifically for tenure and promotion.
3. In all of these institutions, scholarly achievement is expected of all tenure track faculty. This seems a universal requirement. It is what makes universities what they are. Most faculty members would define scholarship as research published in a refereed journal.
4. Finally, university missions of teaching and learning research, extended education, and public service are broader than this narrow view of scholarly achievement. But if a young faculty member comes to a senior faculty member for advice about an opportunity to teach an introductory course with four hundred students while the professor is on sabbatical leave or an opportunity to serve on a multidisciplinary task force on an issue of interest to the citizens of the state but not scientifically oriented, most of us would say the young faculty member would be well advised to wait until tenure is gained.

Oregon State University is a research university and a research grant-oriented university, yet the concept of a broader view of scholarship was readily accepted by a large majority of faculty in the college and later in the university as a whole. Distinguished research professors thought it was a great idea. The Faculty Senate voted unanimously for its adoption. The formidable resistance that people thought existed to viewing scholarship more broadly was imaginary, not real. This assessment is based on Weiser's experiences in presentations during the past years to national science audiences (the American Association for the Advancement of Science, etc.) and at faculties at eighteen universities. In all of these settings and at the workshop, he has found that eighty-five to one hundred percent of the individuals in every group of faculty and university administrators that he has sampled are enthused about accepting a broader view of scholarship (and the resultant implications for changing faculty evaluation and university values) within thirty minutes of discussion. The challenge is to get faculty members to think about the nature of scholarship. Once they do, there seems to be little resistance to adopting a broader value system for evaluating faculty in ways that are congruent with the missions of comprehensive and engaged universities.

The principal premise at Oregon State University was that there was something wrong in any organization or university that has a value system and missions that do not match up. To remedy the situation, a redefinition of scholarship was developed and was used as a basis for revising promotion and tenure criteria and processes. The definition attempted to clearly define scholarship in simple, unambiguous language that applied across all disciplines of a university, and that would be understood beyond academia:

> Scholarship is creative intellectual work that is validated by peers and communicated including the discovery, integration and development of knowledge and creative artistry.

There are a variety of ways to validate scholarship and different ways to communicate it. The definition clearly asserts that scholarship is not only research published in peer-reviewed journals. Oregon State University's view is that scholarship includes the discovery of new knowledge, development or applications of new technology, methods, materials and uses, integration of knowledge leading to new understandings, and creative artistry. Scholarship is done by faculty members engaged in the teaching, research, and outreach missions of a university. The OSU model does not list teaching as a form of scholarship as did Boyer (1992). Scholarship in all areas of faculty work, including research, teaching, and extension is viewed the same as creative work that is peer validated and communicated. Scholarship in teaching was inherently recognized by both Oregon State University and Boyer. But at OSU, validation is focused on the results of scholarship rather than on the process of scholarship as proposed in the Carnegie Foundation report:

- specialty validation of scholarly work is in terms of reproducibility or evidence that something is true or accurate; and
- originality, scope, significance, breadth, depth, duration of influence, impact, public benefit, and usefulness to others.

This recognizes that there are different audiences for different kinds of scholarship and different means of communicating and documenting scholarship. With the electronic and technological developments underway, peer-reviewed journal articles may not remain the predominant avenue for validating and communicating scholarship.

Oregon State University requires that a faculty member's evaluation be based upon their assignment as described in a position description that is reviewed each year. Faculty are evaluated in two primary areas: performance of assigned duties and scholarly achievement. Teaching, research, and outreach is seen as vital university missions and faculty activities, but they are not considered by OSU to be scholarship in themselves unless they involve creative, communicated, peer-validated intellectual work (scholarship). Creative activities such as developing new courses, curricula, and methods for advancing and assessing learning are scholarly contributions when they are shared with others, validated, and communicated.

The pioneering efforts of Oregon State University, Iowa State University, Kent State University, Portland State University, Montclair State University, and the University of Idaho should receive the attention of institutions that wish to revitalize higher education in agriculture. Dean Weiser points out that, in universities, evaluating a faculty member's scholarly contributions and performance of assigned duties is often neither simple nor straightforward, because, in part

- scholarship is undefined and poorly understood at many universities. Scholarship is often simplistically equated with research;
- a faculty member's performance is sometimes evaluated by peers without references to the position description;
- emphasis on individual achievement is sometimes interpreted by peer evaluators to imply that faculty contributions to team efforts are not valuable and important; and
- it is easier to document and evaluate form and activities than substance or consequences.

Comparing the processes that led to change at the aforementioned six universities is also revealing and interesting. For example, the processes at all six universities involved faculty input in a sustained and significant way. In most cases the process was led and carried out by faculty members, although administrators often started the process. The change processes were iterative, took considerable time, and involved lots of discussion and thought. Typically, most took at least one full year. The administrators who were involved characteristically provided encouragement but were not prescriptive regarding outcome. They were generally trusted by their faculties. At all six universities a clear definition of scholarship, broader than the previous view, was adopted. In all cases this broader vision of scholarship provided the conceptual foundation for changes that were subsequently made in faculty evaluation and the tenure and promotion processes. At the six universities where change is occurring, large majorities of faculty and administrators endorsed and accepted the revised evaluation criteria and guidelines for promotion and tenure. A new practice of using a position description as the basis for evaluating a faculty member's performance was adopted by four of the six universities.

Conclusion

Significant change in higher education is occurring in response to the broader visions of scholarship. These systemic changes in what universities value are fundamental cultural changes. Such changes take time. Faculty input is important in establishing broader views of scholarship. New approaches in education often involving the integration of knowledge are not recognized as scholarship in several universities.

Another kind of education of and preparation by faculty members is required. Generally, those scientists who engage in integrative scholarship are inductive thinkers with some years of experience in faculty or government positions or in industrial settings. They may be those who have gained a substantial reputation in the biological or social sciences. They may also be the retirees. Again, as with the substance of values in undergraduate education, scholarship in integration and in education may require more time for intellectual reflection.

There is an undercurrent, however, at a number of colleges of agriculture and natural resources. As institutional fundings become limited, a greater emphasis is being placed on the abilities of faculty members to draw external funding for research that provides the indirect operating (overhead) and partial salary funding that can extend internal funding. This is a trend that has a momentum of its own, often unplanned and not prioritized. External funding is a valuable support of scholarship, to be sure, but it is a value system that may distort the reach for the larger substance of scholarship in an institution.

13

Classroom Environment: Participation

Key Concepts

1. Student participation and interaction in the class space are essential to the learning process and the development of leadership and communication skills.
2. Creating a learning community, not bounded by artificial and self-imposed knowledge limits, should be an instructional goal for each class space.
3. Knowledge is richer when energized by everyone's participation and interaction.
4. It is critical for faculty to learn about the students during the early part of the course so that discussions with them can be productive and meaningful.
5. Participatory education, however, may be difficult to accomplish. Both students and faculty have been indoctrinated in a system of compulsive teaching throughout most of their academic careers.

Introduction

The ideal undergraduate class space extends beyond what most colleges of agriculture perceive it to be. It is the student interaction and their knowledge gain that fix an image of both the instructors and the university. Even before graduation, the class space environment will be reflected as students interact with others: parents, grandparents, siblings, friends, acquaintances, and even casual contacts. It is a particular consequence of the undergraduate classroom that reflections on the instructional context are perhaps the most immediate response of students to the educational process.

Theoretical studies suggest that participatory education is an important element in the implementation of fundamental change in higher education in agriculture, natural resources, and the life sciences. What, for example, might be the effect if the instructor paused in the middle of a class period, after a principle had been presented, and asked the students to pair up and explain the principle to one another? There would likely be a startling and important gain in understanding among the students. At the same time, this simple exercise would clearly bring the student into a participatory mode in his or her education.

Barriers to Participation

Professor James King, University of Nebraska, and Professor Ricardo Salvador, Iowa State University, offered a unique and insightful view of the participatory ambiance in "class space."

An example: A typical science in a soil fertility class for sophomore students. Dose and response to fertilizer are graphically presented. Without fertilizer, plants give some kind of response. There is the point where fertilizer addition results in improved yield. "Facts" are laid out as straightforward facts. Out of nowhere, a student asks a question about economics as it relates to the subject. The professor tersely states, "We are discussing agronomy today and not economics!"

In such a process, student curiosity is squashed; the professor misses educational opportunities. Undue tension may exist in the exchange; the student likely is made to feel inadequate. Communication, the sharing of meaning, between student and professor is destroyed. The class space is bounded by artificial and self-imposed limits of subject matter and knowledge.

Discussion, however, might have related the fertilizer response to many other economic and social aspects. The law of diminishing returns could be taught. The professor might recognize students' concerns. A question is asked, and the professor can find out the answer and cover it in the next class. The opening for student interaction is created: Students who are production minded want to know answers to questions of fertilizer and economic return. An environmentally tuned student will want to know the answer to questions of fertilizer and pollution. Everyone can learn; everyone can teach. In the end, both student and teacher learn from each other.

Students today may be thought of by the faculty as "lumps": they do not read; they are not critical thinkers; they do not talk. But as professors, how are we allowed to criticize student learning when we ourselves do not model inventive and practical learning? In situations like the example above, students are learning much, such as political life, not just narrow disciplines. All students are naturally systems thinkers, but they are conditioned to be nonthinkers in many bounded classes. In real life, all would be disadvantaged if they were not systemic thinkers.

In life, we integrate many things aside from the specific disciplines; thus, it is evident that integrative capacity exists. If the student is a lump and is like the person we criticize, it may be because the student is conditioned to be so by suppressing the integrative capacity in the class.

The Class Space Community

Creating a community of learners should be the goal in every class. The term "learning community" is a meaningful kind of phrase because teaching is about shaping community.

The attitude that the professor carries into the class underpins many of the messages that are sent to the students. For example, activities of the first few days in the course—getting acquainted, letting students know the core values—form the stage for student participation and interaction.

How can a learning community be created? It follows the concept that a learning community is willing to educate one another and to complement one another's weaknesses. In the learning community, we make sure that the knowledge of the entire group is richer because of everyone's participation, instead of the dominance of two or three students in the discussion or of the forced attention to a teacher in monologue. One way of doing that is for faculty to consider the question Who are we? Other questions derived from this question allow students to begin talking about their personal motivation. Suppose we have a student from the Texas Panhandle. The professor asks, "That's an area of increased hog farming. What does your family think of that?" The student replies, "We really don't want the environmental impact of the hogs." "Do you farm?" "Yes." "What does your farm produce?" "Soybeans." "Do you plant only soybeans?" and so on. Next time, the teacher talks with another student.

Consider a class in sustainable agriculture. While the instructor may only be able to talk about it, many students have lived it. Others come from urban areas or large industrialized farms. They are interested in the sustainable business too. They want to learn the language and the concepts. They want to know what the argument is all about. Thus, it is possible to know people more as humans rather than just the rigor of their sharp scientific intellect.

An additional technique is to ask the students to be responsible for being ready to answer questions in their own particular subject area. Part of the class period can be devoted to students delivering their own subject matter within the context of the course material.

Why Students Do Not Participate

The perceptions of students and teachers are very different as to why some students do not speak up. Research says there are three reasons for this. First, students believe that they can tell very early in the semester whether or not the professor really wants the students to prepare for class. Students will also be aware that a concentration of responsibility exists: only a small group of students can be counted on to respond to questions or have comments on an issue raised in class. Second, the students feel fairly safe that they will be rarely called on when the teacher calls on specific students. Third, students are tested infrequently.

Student Participation and Skills

King and Salvador argue that student participation and interaction is vital to the learning process and the development of leadership and communications skills. Persons addicted to being taught seek security in compulsive teaching. Persons who experience knowledge as a result of a process want to reproduce it in others. It is necessary to instill a sense of responsibility in students so that they can be leaders in the transformation to life-long learning. Participation is key to the development of critical thinking and teamwork skills that will be needed to solve the complex problems facing society.

Recent developments in technology and pedagogy have made available a wide range of new possibilities for facilitating the learning process, but they also create new challenges for effective student interaction. For example, the Internet, compressed video, and CD-ROMs offer informational and content enhancements to students and help accomplish goals of integrated education, leadership development, and student participation. Technology, however, is costly and requires an investment of teacher energy and time. King and Salvador emphasize the need of faculty to use innovation and new technological, pedagogical, and curricular approaches that will allow students to cooperate in problem-focused, multidisciplinary, educational environments.

Challenges exist: encouraging students to take responsibility and an active role in the learning process; providing incentives for faculty to innovate; evaluating young faculty's innovation; and training young faculty to encourage student participation.

Conclusion

Students, themselves, can be and ought to be engaged in their own education. To do that may not be just allowing the student a wider selection of courses in his or her course of study, although that has value. Teachers will have to take on more complex roles in the class space and laboratories, particularly in the expanded electronic class space. They will have to become less authority figures whose intent is to deliver facts (which the future may question as facts) and become more colearners and facilitators whose purpose is to encourage questioning, exploration, and synthesis by their students. Participation by everyone involved in the class space environment should provide all who want to learn with access to the wonderful resources of higher education.

Students and faculty, however, are largely secure in the teaching system generally used. They have been subject to such a system in most of the courses they had in high school and college. Faculty members like to lecture. Thus, the change to participatory education, as with problem-based learning, may be difficult for some faculty members to adopt.

Faculty members can change even though they have been caught up in a system of teaching dominant in colleges of agriculture for decades, but time, effort, and a satisfactory experience will be required. Participatory learning may be the one way that critical thinking can be taught, and thus its value should be unquestioned.

14

Innovation and Systems Thinking

Key Concepts

1. Innovation consistent with the recommendation concerning systemic change is already occurring in colleges of agriculture and natural resources. However, most educational reforms originate in isolated classrooms, institutions, and professional societies. A new system is required to institutionalize such individual efforts.
2. Systems thinking can be applied to the educational process as well as be a concept integrated in course and curricular contents. Integrated education can be achieved aided by technological advances such as flexible and updated digital presentation, Web-based education, interconnection of courses taught in the same semester or quarter, and cooperative education. Such education can integrate societal and ethical issues into a framework of scientific and technical education.
3. Relevance of course contents and evident interconnections among them are important to the student.
4. Technical competence is an important outcome of higher education. But perception of the whole agricultural or natural resources system and the ability to communicate are also essential for graduates of agricultural and natural resource curricula. Process thinking should replace functional thinking as attributes of graduates in agriculture and the natural resources.

Introduction

Reform efforts in higher education in agriculture and natural resources are actually occurring in several colleges of agriculture, but they originate as isolated cases. As faculty members and others discuss the issues of change and implementation, their thoughts will often approach the matter as an individually related experience: the faculty member most dedicated to change will likely fashion changes within their own teaching paradigm and setting.

Such individual experiences, however, can be useful as ideas for others if communicated. Individual institutional experiences can accumulate impacts. They can lead to some general guidelines for broad institutional implementation of change. The workshops included such examples.

Timothy Mack, professor of entomology, Virginia Polytechnic Institute and State University, presented a case in which the college of agriculture provided for general education and developed certain interinstitutional networks. Joe Colletti, associate professor of forestry, Iowa State University, elaborated on the leadership of the faculty in forestry in perceptually involving students with the faculty to further their own learning toward new areas. Principles of innovation emerged from each of the discussions.

Doing More (with Less?)

Mack's discussion began with the description of the current environment in colleges of agriculture. There is an emphasis among college of agriculture and natural resources faculties on gaining computer knowledge that may be displacing the emphasis on gaining discipline-specific knowledge. Faculty members are expected to become more involved in the undergraduate experience while the number of faculty members is decreasing.

Technology now has massive effects in the area of higher education. As a result, new for-profit institutions may be skimming off profitable aspects of education. Their targets, however, will not likely be elements in higher education in agriculture and natural resources, but rather those elements that form core structure in universities and colleges. Revitalized colleges of agriculture and natural resources can and will remain largely residential institutions, but public institutions will be linked with for-profit institutions in that students already transfer credit to the public university, bringing their education with them.

Mack noted that in the current technological world, a set of digital (PowerPoint, CD-ROM) presentations can be packaged and sold as a text book. A digital presentation can be created in which multiple professors can have impacts. The classes taught then can be in small discussion groups (the students have the slides), in large classrooms, and in a network of institutions. Such presentations can be put into frameworks understood by students of diverse backgrounds and goals.

Mack illustrated his points by telling about a specific course in general entomology that now draws hundreds of students. Relevance is gained by talking about historical impacts, such as the plague that killed over one-third of the world's population in 1348–1350. It is gained by talk about insects in art museums and in literature. There was discussion on arachnophobia (*Charlotte's Web*). An insect-borne disease, malaria, is the leading disease in the world. And, so on. As a further development, the Department of Entomology at Virginia Polytechnic Institute and State University took the PowerPoint presentations and converted them to Web pages. Snippets of on-campus lectures with real audio were incorporated. Whoever has a computer, modem, and Internet Explorer or Netscape configured properly should be able to access the presentations. The enrollment in the distance learning class is actually growing at a faster rate than the on-campus class.

However, distance learning in such a setting is expensive. Mack estimated the costs at a million dollars per course, in faculty salaries, equipment costs, consultants, programmers, and facilitators. A fundamental philosophy is that the Web-based class should have some interactivity for active learning. Online submission of assignments is available. Testing is difficult because of the need to monitor. Mack suggested that the extension county agents may be available for a local access network for proctoring and registration. To accomplish that, agreements would have to be reached and funding patterns changed. An electronic means of questioning and responding has been devised. Mack states, "There is a fallacy to think that distance education does not have interaction with students. What is missing is the face to face contact." But, from the student perspective, is that different from being in a large class? The question was left unanswered.

Problems in such development still have to be worked out. Further logistic testing is needed. Working with other universities, each with its own registration and tuition structure may be difficult but can be explored. Ownership of the credit and tuition fees is a problem. Obviously, administrative support is needed.

Evident in the case that Mack described is the opportunity of a college of agriculture and/or natural resources to make a significant, unique contribution to the general education of students. As it develops, technology may enable the college to do more with the resources it has. Faculty skills in such technologically based instruction are likely to increase in the future.

But, regardless of the sophistication of the technology, content remains the vital component.

Doing More Together

Joe Colletti, associate professor of forestry, Iowa State University, presented a case of evolution in the educational philosophy from one in which the knowledge is presented to students regardless of whether the individual student absorbed it to a more student-centered learning. In the old curriculum, a summer camp was expected to provide the experiential base. In the new curriculum, lectures are still given, not minimized. The faculty still want the students to have experience, but the intent is to be more engaging with the students. As the typical requirement for the summer camp was removed from the program, the question was posed, What can we do to facilitate the higher level of understanding and learning in our students?

First, the department at Iowa State University knew that the students should be technically competent. But feedback from the profession was that students, though technically competent, did not know how to work well in teams in industry and governmental jobs, a necessity for the future. Thus, the department has gone to much more interactive and team-oriented actions.

A key point, as Colletti described it, was how to link interactive and team skills with other courses in the curriculum. If students do things and verbalize things, they are going to achieve enhanced learning. A tool in all courses, from freshman to senior level, is to facilitate learning from synthesis to analysis and evaluation, not just acquisition of basic knowledge. The department wants the students to integrate social, economic, and environmental factors as well as institutional and political factors in complex decision making. This thought process changed the way learning in forestry was approached.

The students in the Forestry Program at Iowa State University have a set of core curriculum requirements: Nine and a half credits of interpersonal public communication skills, seventeen credits of mathematics and physical and life sciences, and fifteen credits in humanities, ethics, and social sciences. Additionally, students take at least three credit hours of communication-intensive, environmental-intensive, and problem solving-intensive courses within forestry. An association has occurred with the Colleges of Engineering and Education with a focus on student learning: "How can we more effectively incorporate student-centered learning into our curriculum?" Faculty have engaged in workshops and biweekly groups as part of Project LEA/RN (Learning Enhancement Action/Resource Network) to learn about modern learning strategies and classroom assessment techniques.

The unique approach described by Colletti is that classroom arrangements need to be flexible, such as having moveable tables instead of bolted down chairs. It is a matter of thinking differently about the environment in which interactive student learning has to occur. The need exists for computers that facilitate visualization and the communications that are expected of the students, especially from the sophomore to senior level. A new array of technological aids has been made available. But a philosophical change is also occurring in the college climate: encouragement of thinking about issues in a broader sense and of thinking about process in addition to function, communication, and adaptation.

15

Strategies for Implementation

A widely expressed theme in higher education in agriculture is that the system should produce "society-ready graduates" (Carpenter and Fisher, 1996).

To do so, land grant and other universities offering higher education in agriculture and natural resources must be in touch with the needs and expectations of society. We agree. But we conclude from these studies supported by the systemic change projects that more is required than dialogue between the college and potential employers, although the importance of that is unquestioned. What is also required is an understanding of the kind of society and industry in which graduates will work in the future by administrators, faculty, and clientele of the academy. The essential requirement to implement change is to see the nature of the world and our environment well into the twenty-first century. Society is in a new trajectory. What is happening may not be easily described, but the next requirement in revitalizing higher education in agriculture and natural resources is scholarship that can describe the future needs of society.

We do know that the agrarian age has passed in the United States and in most developed countries. We know that an industrial society followed, providing stability for college graduates, their work, and their living environments. That too is passing, although higher education in agriculture and natural resources is still largely predicated on the industrial society. The traditional societies for agriculture are being replaced by a society that might be defined as a risk society with global dimensions.

Graduates of the future will likely face a number of changes in their careers, not necessarily in the pattern of "climbing the corporate ladder." Lateral job movement will be as likely as upward movement. The graduates will likely live in multiple locales.

Significantly, society is viewing agriculture, food, and natural resources in terms that are different from those through most of the twentieth century. Society, through its governmental and legislative routes, may place even greater restrictions on certain aspects of agriculture and natural resources usage because it views the risks of those activities to be too great. European and some third world countries and environmentalists throughout the world are attempting to gain world concurrence to restrict trade in genetically modified plants and animals and their products, changing the market for United States agriculture. However, society, with its concerns about food safety, genetically modified organisms, and other issues, may be beginning to sense the importance of food and agricultural sciences in their lives.

Thus, the most important step in implementing the needed fundamental changes in higher education in agriculture and the natural resources is to gain a perceptive understanding of the society and the industries for which our graduates must be readied. Scholarship to gain these insights now becomes a necessary strategy to implement the needed changes.

What is this new needed perception of the world? We can begin to have some concept of what it will be. For one thing, we know that society is in transition, particularly as it relates to the food system. Human health and welfare are the reasons for the agricultural system. Paul Thompson suggested that the description of a risk society by the author Ulrick Beck may be applicable. While the old issues of How am I going to stay in business? remain, issues that affect all individuals are demanding attention: food safety, human nutrition, which even now goes beyond knowledge of required nutrients, air and water pollution, resource depletion, land use, climate change, animal welfare, dietary rights, race and gender discrimination, property rights, and quality of rural life are some of the subject matters of revitalized colleges of agriculture. Although describing the world today as a risk society may be going further than we are prepared to go at this time, the concepts and contexts of a risk society ring well with us. Our recent experience is that graduates of animal science at Texas A&M University who receive early offers of employment are those who are able to take employment in some area of quality control.

The questions are obvious:

- Are students prepared for a society that is rapidly changing? Are they thinking globally?
- Are faculties prepared to teach them?

Students, themselves, bear risk of being prepared for changes in jobs, homes, and partners. Are they prepared to face these risks? They will also have to deal with the questions of risk that are asked about their work and be able to cope with them.

These studies have pointed to the primary needs, that newly minted graduates have a certain level of technical competence and the ability to communicate as well as a relevant background knowledge. The ability to communicate requires not only writing and speaking skills, but the ability to perceive and solve problems. Graduates in their first job then need the ability to develop the skills and knowledge base needed in their subsequent careers.

Facing this, colleges may *implement* the needed changes by

1. Engaging reality by describing themselves as institutions concerned with biologically and ecologically based systems and the relevant social sciences for the benefit of humans and natural populations.
2. Rethinking their goals, values, and missions and providing the necessary resources for excellence in undergraduate and graduate studies.
3. Recognizing that economic activity in agriculture in the future will be knowledge based. High technology will likely dominate the increased interaction between society and industry, and with it will come greater concern for risk and regulation and intellectual property. Such concepts will drive graduate education in colleges of agriculture and natural resources and should become part of the conceptual basis for curricular and course development for undergraduate education.

   The future undergraduate education in agriculture and natural resources must not be aimed solely at the initial employment. It should be for the long-range trend determined by unstable societal/economic/environmental perspectives.
4. Accepting the principle that implementation of systemic change in higher education in agriculture and the natural resources is a shared responsibility of administration, faculty, and students.
5. Creating the climate for change by defining appropriate insights as well as the concepts of higher education for the twenty-first century.
6. Providing opportunity for moral talk.
7. Constructing courses and curricula that are outcome, not content, driven and providing opportunities within those curricula and courses for students to learn both technical knowledge and the skills of communication and problem solving.
8. Devising course and curricular contents to provide fundamental knowledge needed for twenty-first century careers and employing the faculty or restructuring faculty assignments to be able to teach such courses.
9. Providing greater recognition and opportunity for integrative and educational scholarships as well as discovery and application as incentives for faculty tenure, promotions, and salary increases.
10. Providing greater opportunity for students to participate in their own education.
11. Using technological development to incorporate systems thinking in the process of education as well as a content of the education.

The strategies are not alternatives to one another but we suggest that they all be tested as part of the fabric of change in higher education in agriculture and natural resources to develop an education that has relevance in the decades ahead. At least, administrations and faculty need to take time to think about such matters.

In revolutionizing higher education in agriculture, however, some of the old values will remain intact. Skill in instruction will remain a highly valued attribute of faculty members. Faculty members that teach well and maintain rigor in their courses are universally needed in higher education. Higher expectations of all students are important.

These studies on implementation of change came primarily out of two workshops—actually, focus group discussions—in which the direction was set by the core Work Group on the Implementation of Systemic Change in Undergraduate Education in Agriculture. The speakers, however, generally took free rein in what they said, and the discussions by the participants were open and widely ranging. The workshops were, in fact, creative and served the objectives of the projects on systemic change in higher education in agriculture and the natural resources. This synthesis is the report of the projects, and the recommendations and suggestions represent only some of the solutions to the needed fundamental changes and their implementation. We believe that if they are brought into discussion among faculties of agriculture and natural resources, substantial progress will be made in revitalizing higher education in agriculture.

Epilogue

The epilogue of *Revolutionizing Higher Education in Agriculture* (Kunkel et al., 1996) began with the statement, "This study is neither complete nor final." It ended with the imperative, "Envisioning boldly, determining values, setting priorities, taking risks, but retaining concern for the human aspects of higher education—these are the elements of effective development of higher education in agriculture and all its related area of knowledge."

Noted also was that "new rules of agriculture, business, commerce and life are being written. . . . In matters as serious as these, the genesis of new ideas and new ways of putting things together must always be welcome."

With the fresh look at strategies for implementation of systemic change, the concept is further transformed by consideration of what really occurs in the educational process. There is teaching and learning of theories, facts, formulas, procedures, and other routine activities. A subtle but persuasive activity is also occurring. In the process of doing and teaching the sciences and technologies, people are teaching and learning "ways of seeing" and "ways of being" (Grinnell, 1987, cited by Burkhardt). "Teachers impart and students imbibe appropriate technical lessons; they also instruct and internalize other perceptional and attitudinal norms."

We urged the educational community to act boldly. We encouraged them to seek and create new ideas of how to put things together. Now we take our clue from what actually happens and see that bold action and new ideas are implemented by the ways of seeing and being that are changed. Inherent in the ability to change, incrementally or systemically, is the individual and institutional conception, that is, that agriculture and natural resources have multiple functions in contemporary society, that one does not need to teach everything in a course, that the course be goal driven, perhaps that the work environment in which we place our students is a risk society, that students must participate in their own education, that our teaching can utilize a systems approach, or that the system should reward integrative scholarship as much as it does science.

What a difference it would make if these ways of seeing and being were integrated with boldness, incentives, and working together. Once again, we state it: scholarship to gain these insights now becomes a necessary strategy to revolutionize higher education in agriculture and the natural resources.

Appendix

First Workshop: College Station, Texas, 1–3 May 1997

Presentations

Welcome: E. A. Hiler, Agriculture and Life Sciences, Texas A&M University System

Concepts of implementation and change: H. O. Kunkel, Texas A&M University

Integration and Systems Thinking: Stuart Gage, Michigan State University

Values in Undergraduate Programs: Jeffrey Burkhardt, University of Florida

Innovation and Student Participation: James King, University of Nebraska; Ricardo Salvador, Iowa State University

Administrative Support: McArthur Floyd, Alabama A&M University; Suman Singha, University of Connecticut; H. Dean Sutphin, Cornell University; A. Gene Nelson, Texas A&M University, Moderator

Participants

Horace H. Baily, West Texas A&M University, WTAMU Box 267, Canyon, TX 79016

Richard Barrows, University of Wisconsin-Madison, 1450 Linden Drive, Madison, WI 53706

Arlo Biere, Kansas State University, 117 Waters Hall, Manhattan, KS 66506-4015

Peter Burfening, Animal and Range Sciences, Montana State University, Bozeman, MT 59717

Jeffrey Burkhardt, Food and Resource Economics, University of Florida, Gainesville, FL 32611

Jimmy G. Cheek, University of Florida, Gainesville, FL 32611-0270

James Clark, West Texas A&M University, WTAMU Box 998, Canyon, TX 79016-0001

Robert Coulson, Entomology, Texas A&M University, College Station, TX 77843-2475

C. R. Creger, Executive Associate Dean, College of Agriculture and Life Sciences, Texas A&M University, College Station, TX 77843-2142

Larry H. Erpelding, Kansas State University, 117 Waters Hall, Manhattan, KS 66506-4015

McArthur Floyd, Associate Dean for Research, School of Agricultural and Environmental Sciences, Alabama A&M University, Normal, AL 35762

Stuart Gage, Entomology, Michigan State University, East Lansing, MI 48824

James Gilley, Agricultural Engineering, Texas A&M University, College Station, TX 77843-2117

Aurora S. Hodgson, University of Hawaii at Manoa, 1800 East-West Road, Henke 224, Honolulu, HI 96822

Bryan H. Johnson, Animal Science, Texas A&M University, College Station, TX 77843-2471

Bob L. Karr, Mississippi State University, Box 9680, Mississippi State, MS 39762

Ann L. Kenimer, Agricultural Engineering, Texas A&M University, 201 Scoates Hall, College Station, TX 77843-2117

Jim W. King, Agricultural Leadership, Education and Communication, University of Nebraska, Lincoln, NE 68583

Kendrea Kirkpatrick, Student, Agricultural Education, Texas A&M University, College Station, TX 77843

LeRoy Klein, Schools of Medicine and Engineering, Case Western Reserve University, Cleveland, OH 44106-5043

Duane C. Kraemer, College of Veterinary Medicine, Texas A&M University, Room 101 VMA, College Station, TX 77843-4461

Karen Kubena, Associate Dean, College of Agriculture and Life Sciences, Texas A&M University, College Station, TX 77843-2402

H. O. Kunkel, Animal Science, Texas A&M University, College Station, TX 77843-2471

Don Lindsey, New Mexico State University, P.O. Box 30003, Dept. 3BE, Las Cruces, NM 88003

Kerry K. Litzenberg, Agricultural Economics, Texas A&M University, College Station, TX 77843-2124

Rebecca Lochmann, University of Arkansas-Pine Bluff, P.O. Box 4912, Pine Bluff, AR 71611

Bill Mautz, Dean, College of Agriculture and Life Sciences, University of New Hampshire, Taylor Hall, Durham, NH 03824

Richard McNeil, Natural Resources, Cornell University, Fernow Hall, Ithaca, NY 14853

Brooke Miller, Student, Animal Science, Texas A&M University, College Station, TX 77843

Murray Milford, Soil and Crop Sciences, Texas A&M University, College Station, TX 77843-2474

Mark Murphey, Student, Animal Science, Texas A&M University, College Station, TX 77843

William A. Neill, Wildlife and Fisheries, Texas A&M University, College Station, TX 77843-2258

A. Gene Nelson, Agricultural Economics, Texas A&M University, 333 Blocker Building, College Station, TX 77843-2124

Cathy A. Paris, Botany and Agricultural Biochemistry, University of Vermont, Burlington, VT 02405-0086

Patricia Pietrantonio, Entomology, Texas A&M University, College Station, TX 77843-2475

Manuel Pina, Agricultural Education, Texas A&M University, College Station, TX 77843-2116

Ricardo Salvador, Agronomy, Iowa State University, Ames, IA 50011

Alan Sams, Poultry Science, Texas A&M University, College Station, TX 77843-2472

Howard Sandberg, HEP, SERD, CRS, EES, USDA, Fourteenth and Independence Avenue SW, Washington, DC 20250-2251

Suman Singha, Associate Dean, Academic Programs, College of Agriculture, University of Connecticut, 1380 Storrs Road, Storrs, CT 06269-4090

Chris L. Skaggs, Animal Science, Texas A&M University, College Station, TX 77843-2471

Grant Suhm, Agricultural Education, Texas A&M University, College Station, TX 77843-2116

Dean Sutphin, Associate Dean, College of Agriculture and Life Sciences, Cornell University, Roberts Hall 140, Ithaca, NY 14853-5901

Pete D. Teel, Entomology, Texas A&M University, College Station, TX 77843-2471

Paul B. Thompson, Philosophy & Humanities, Texas A&M University, College Station, TX 77843-2472

Donald Vietor, Soil & Crop Sciences, Texas A&M University, College Station, TX 77843-2474

Robert Whitson, Rangeland Ecology and Management, Texas A&M University, College Station, TX 77843-2126

Second Workshop: Ithaca, New York, 23–25 April 1998

Presentations

Welcome: Daryl B. Lund, Agriculture and Life Sciences, Cornell University

The Workshop and the Process: H. O. Kunkel, Texas A&M University

Course Design: George Posner, Cornell University

Course Content: Paul B. Thompson, Purdue University
Scholarship and Teaching: C. J. Weiser, Oregon State University
Innovation, Experiential Learning, and Networking: Timothy Mack, Virginia Polytechnic Institute and State University; Joseph Colletti, Iowa State University
Synthesis of Strategies: Participants

Participants

David S. Barrington, Department Chair, University of Vermont, 225B Marsh Life Sciences Building, Burlington, VT 05405
Perry Clark, Chair, Agriculture and Environmental Science, University of Wisconsin-River Falls, 410 South Third, River Falls, WI 54022
Joseph P. Colletti, Forestry, Iowa State University, 243 Bessey Hall, Ames, IA 50011-1021
Terry Ferries, Associate Dean, College of Agriculture, Food and Environmental Science, University of Wisconsin-River Falls, 410 South Third, River Falls, WI 54022
McArthur Floyd, Associate Dean for Research, Alabama A&M University, P.O. Box 1087, Normal, AL 35762
LeRoy X. Klein, Biochemistry and Macromolecular Science, Case Western Reserve University, 11100 Euclid Avenue, Cleveland, OH 44106
H. O. Kunkel, Dean Emeritus, Agriculture and Life Sciences, Texas A&M University, College Station, TX 77843-2471
Daryl B. Lund, Dean, College of Agriculture and Life Sciences, Cornell University, 260 Roberts Hall, Ithaca, NY 14853-5901
Timothy P. Mack, Head, Entomology, Virginia Polytechnic Institute and State University, Blacksburg, VA 24061-0319
Richard McNeil, Natural Resources, Cornell University, 106 Fernaw Hall, Ithaca, NY 14853-3001
John Nicholaides, University of Illinois, 1102 South Goodwin Avenue, Urbana, IL 61801
Susan C. Piliero, Educaton, Cornell University, 414 Kennedy Hall, Ithaca, NY 14853
E. John Pollak, Animal Science, Cornell University, B-22 Morrison Hall, Ithaca, NY 14853
John Riesen, Animal Science, University of Connecticut, 3636 Horsebarn Road, Storrs, CT 06269-4040
Suman Singha, Associate Dean, Academic Programs, College of Agriculture, University of Connecticut, Storrs, CT 06269-4090
Chris Skaggs, Animal Science, Texas A&M University, College Station, TX 77843-2471

James R. Strickland, Nutritional Toxicology, New Mexico State University, Box 30003, MSC 3-1, Las Cruces, NM 88002

H. Dean Sutphin, Associate Dean and Director of Academic Programs, College of Agriculture and Life Sciences, Cornell University, 140 Roberts Hall, Ithaca, NY 14853-3081

Paul B. Thompson, Philosophy, Purdue University, West Lafayette, IN 47907-1306

Donald R. Viande, Associate Director of Academic Programs, College of Agriculture and Life Sciences, Cornell University, 140 Roberts Hall, Ithaca, NY 14853-5901

Donald M. Vietor, Soil and Crop Sciences, Texas A&M University, College Station, TX 77843-2474

C.J. Weiser, Dean Emeritus, College of Agricultural Sciences, Oregon State University, Corvallis, OR 97331

Tryon A. Wickersham, Student, Animal Science, Texas A&M University, 5950 Hearne Road, Bryan, TX 77808

Edward L. Williams, Alabama A&M University, P.O. Box 1087, Normal, AL 35762

M. Karl Wood, Animal Sciences, New Mexico State University, Box 3-1, Las Cruces, NM 88003

References

Alexander, J. C. 1993. The irrational disciplinarity of undergraduate education. Chronicle of Higher Education, p. B3.

American Vocational Association, Agricultural Education Division. 1989. Strategic plan for agricultural education. National Summit on Agricultural Education, Washington, DC.

Anderson, R. E., and J. W. Meyerson, eds. 1992. Productivity and Higher Education: Improving the Effectiveness of Faculty, Facilities, and Financial Resources. Princeton, NJ: Peterson's Guides.

Barinagar, M. 1994. Agricultural research: A bold new program at Berkeley runs into trouble. Science 263:1367–1368.

Barry, P. J., J. R. Evans, L. Otto, J. W. Richardson, and J. Schultz. 1994. The social sciences: Research programs, interdisciplinary work, and communications. ESCOP Social Sciences Committee, National Association of State Universities and Land Grant Colleges, June.

Bass, B. M. 1975. Leadership Psychology and Organized Behavior. Westport, CT: Greenwood Press.

Beattie, B. R. 1983. Ten commandments for agricultural economics (and perhaps other) administrators. Western Journal of Agricultural Economics 8:209–214.

———. 1991. Some almost-ideal remedies for healing land grant universities. American Journal of Agricultural Economics 73(5):1307–1321.

Beck, U. 1992. Society: Towards a Modernity. Trans. M. Ritter. London: Sage Publications.

Benningson, L. A., and H. Swartz. 1987. The CEOs change agenda. Planning Review (3):13.

Berg, A. 1993. Sliding toward nutrition malpractice: Time to reconsider and redeploy. American Journal of Clinical Nutrition 57:3–7.

Bierzychudek, P., and C. G. Reinnes. 1992. Helping nonmajors find out what's so interesting about biology. BioScience 42:125–128.

Bonnen, J. T. 1986. A century of science and agriculture: Lessons for science policy. American Journal of Agricultural Economics 68:1063–1080.

Boyer, E. 1990. Scholarship Reconsidered: Priorities of the Professorate. Princeton, NJ: Carnegie Foundation for Advancement of Teaching.

Bradshaw, D. E., and D. J. Marquart. 1990. New age professionals for a new agriculture. Agrichemical Age, pp. 24–25.

Brentlinger, W. B. 1986. Curricula for the twenty-first century. Paper presented at Regional Conference on University Teaching, Las Cruces, NM.

Bromley, D. W. 1992. Vested interests, organizational inertia, and market shares: A commentary on academic obsolescence. Choices 7(3): 58–59.

Bullock, B., N. Clarke, T. Dutson, C. Krueger, and C. Scifres. 1993. External and internal environments of the state agricultural experiment stations. Towards the Twenty-first Century: A Multidimensional Transition of the State Agricultural Experiment Stations. National Futuring Conference, Experiment Station Committee on Organization and Policy, 15–16 June, Washington, DC, pp. 5–11.

Busch, L., and W. B. Lacy. 1983. Science, Agriculture, and the Politics of Research. Boulder, CO: Westview Press.

Byé, P., and M. Fonte. 1993. Toward science-based techniques in agriculture. Agriculture and Human Values 10(2):16–25.

Cardwell, V. B. 1985. Principles approach in teaching crop management courses. Journal of Agronomic Education 14:98–103.

Carmichael, J. W., Jr., D. D. Labat, J. T. Hunter, J. A. Privett, and J. O. Sevenair. 1993. Minorities in the biological sciences: The Xavier success story and some implications. BioScience 43:564–569.

Carpenter, Z. L., and J. R. Fisher. 1996. From Issues to Action: A Plan of Action on Agriculture and the Natural Resources for Land Grant Universities. Washington, DC: National Association of State Universities and Land Grant Colleges.

Carter, J. L., E. Heppner, R. H. Saigo, G. Twitty, and D. Walker. 1990. The state of the biology major. Bioscience 40 (9): 678–683.

Castle, E. N. 1980. Agricultural Education and Research: Academic Crown Jewels or Country Cousin? Washington, DC: Resources for the Future.

———. 1991. Preface to Proceedings of the Fourth Annual Meeting of the National Rural Studies Committee. Western Rural Development Center, Oregon State University.

Cohen, W. A. 1990. The Art of the Leader. Englewood Cliffs, NJ: Prentice Hall.

Collins, J. C., and J. I. Porras. 1991. Organizational vision and visionary organization. California Management Review 34(1):30–52.

Coulter, K. J. 1994. Personal communication.

Council of Administrative Heads of Agriculture (CAHA) Committee on Public Lands. 1993. Unpublished document.

Creswell, J. W., D. W. Wheeler, A. T. Seagren, N. J. Egly, and K. D. Beyer. 1990. The Academic Chairperson's Handbook. Lincoln, NE: University of Nebraska Press.

Curry, B. K. 1992. Instituting enduring innovations: Achieving continuity of change in higher education. Report 7, 1992, ASHE-ERIC Higher Education Report Series.

Danbon, D. B. 1979. The Resisted Revolution: Urban American and the Industrialization of Agriculture. Ames, IA: Iowa State University Press.

Derbertin, D. L. 1992. There is a future for land grants, if . . . Choices 7(3):47.

Doering, O. C., III. 1992. The social and ethical context of agriculture: Is it there and can we teach it? In National Research Council, Agriculture and the Undergraduate. Washington, DC: National Academy Press, pp. 237–242.

Donnelley, S., C. R. McCarthy, and R. Singleton Jr. 1994. The brave new world of animal biotechnology. Hastings Center Report 24 (Special Supplement 1):S1–S31.

Drucker, P. F. 1994. The age of social transformation. The Atlantic Monthly, November, pp. 53–80.

———. 1999. Beyond the information revolution. The Atlantic Monthly, October, pp. 47–57.

Eisen, A., J. G. Morgan, and P. A. Marstellar. 1992. Methods of teaching investigative skills in college introductory biology—pilot to program. BioScience 42:870–873.

Elliott, C. 1992. Where ethics come from and what to do about it. Hastings Center Report, July–August, pp. 28–35.

Esmay, M. L. 1991. Ethics of U.S. agricultural research in international development. In Beyond the Large Farm: Ethics and Research Goals for Agriculture, ed. P. B. Thompson and B. A. Stout. Boulder, CO: Westview Press, pp. 161–174.

Farmer, D. W. 1988. Enhancing student learning: Emphasizing essential competencies in academic programs. Kings College, Wilkes-Barre, PA. Monograph.

———. 1990. Strategies for change. In Managing Change in Higher Education. New Directions in Higher Education, no. 71, ed. D. W. Steeples. San Francisco: Jossey-Bass, pp. 7–18.

Fife, J. D. 1992. Forward to Instituting enduring innovations: Achieving continuity of change in higher education. Ed. B. K. Curry. Report 7, 1992, ASHE-ERIC Higher Education Report Series.

Food and Agriculture Education Information System (FAEIS). 1999. Fall 1998 enrollment for agriculture, renewable natural resources, and forestry. Food and Agriculture Education Information System, College Station, TX.

Gibson, C. C. 1993. Teaming up for teaching and learning: Redefining the team and the field of play. Unpublished paper presented at the Ninth Annual Conference on Teaching and Learning at a Distance, August 4–6, Madison, WI.

Glassick, C. E., M. T. Huber, and G. I. Maeroff. 1997. Scholarship Assessed: Evaluation of the Professoriate. San Francisco: Jossey-Bass.

Goldberg, R. A. 1993. New international linkages shaping the U.S. food system. Choices, 4th Quarter, pp. 15–17.

Goodman, D., B. Sorj, and J. Wilkinson. 1987. From Farming to Biotechnology. Oxford, England: Basil Blackwell.

Gordon, J. C. 1992. The environmental curriculum: An undergraduate land grant future? In National Research Council, Agriculture and the Undergraduate. Washington, DC: National Academy Press, pp. 51–54.

Grinnell, F. 1987. The Scientific Attitude. Boulder, CO: Westview Press.

Handelsman, J. 1992. Changing the image of agriculture through curriculum innovation. In National Research Council, Agriculture and the Undergraduate. Washington, DC: National Academy Press, pp. 199–203.

Harl, N. E. 1993. A potpourri of ideas on undergraduate education. Review of the book Agriculture and the Undergraduate. Journal of Agricultural Economics Research 44(4):41–44.

Hartel, P. G., K. P. George, and J. Vorst, eds. 1994. Agricultural ethics: Issues for the twenty-first century. Special Publication no. 57. American Society of Agronomy, Crop Science Society of America, and Soil Science Society of America, Madison, WI.

Hefferlin, J. B. L. 1972. Hauling academic trunks. In Elements Involved in Academic Change, ed. C. U. Walker. Washington, DC: Association of American Colleges, pp. 1–10.

Heichel, G. H. 1990. Communicating the agricultural research agenda: Implications for policy. Journal of Production Agriculture 3:20–24.

Hershey, H. G. 1986. The impediments to curricular change. Journal of Dental Education 50:102–106.

Holt, J. 1989. Managing change in extension. American Journal of Agricultural Economics 71:869–873.

Jick, T. D. 1993. Managing Change: Cases and Concepts. Homewood, IL: Richard D. Irwin.

Jones, R. L. 1989. Sunrise legacy in place at college of agriculture. Advance 1:1, 6.

Kanter, S. 1992. The buck stops here: Outside grants and the general education curriculum change process. Working paper no. 9. Exxon Education Foundation, New York.

Kellogg Commission. 1997. Food Systems Profession Education Initiative. Washington, DC: National Association of State Universities and Land Grant Colleges.

Kellogg, C. E., and D. C. Knapp. 1966. The College of Agriculture: Science in the Public Service. New York: McGraw-Hill.

Kinsella, J. E. 1993. Strategic plan for the college of agricultural and environmental sciences. University of California, Davis.

Kirkendall, R. S. 1986. In Agriculture and Education, ed. G. C. Fite. Washington, DC: Agricultural History Society, pp. 3–21.

Kirkpatrick, D. L. 1985. How to Manage Change Effectively. San Francisco: Jossey-Bass.

Kolb, D. 1984. Experiential Learning: Experience as the Source of Learning and Development. Englewood Cliffs, NJ: Prentice Hall.

Korn, H. 1986. Psychological Models Explaining the Impact of College on Students. Ann Arbor, MI: National Center for Research to Improve Postsecondary Teaching and Learning, University of Michigan.

Kunkel, H. O. 1992. The changing mandate of colleges of agriculture: A national committee assessment. Journal of Natural Resources and Life Sciences Education 21:120–122.

———. 1997. Integration of research: Theoretical perspectives and practical considerations. Cells and Materials 7:271–277.

Kunkel, H. O., and G. R. Hagevoort. 1994. Construction of science for animal agriculture. Journal of Animal Science 72:247–253.

Kunkel, H. O., I. L. Maw, and C. L. Skaggs. 1996. Revolutionizing Higher Education in Agriculture: Framework for Change. Ames, IA: Iowa State University Press.

Lindsey, P. J., and M. V. Martin. 1993. The agricultural business management program: Lessons learned at Oregon State University. Journal of Agribusiness, 11–2 (Fall):101–116.

Litzenberg, K. K., and V. E. Schneider. 1987. Competencies and qualities for agricultural economics graduates sought by agribusiness employers. American Journal of Agricultural Economics 69(5): 1031–1036.

Madden, J. P. 1986. Toward a new covenant for agricultural academe. In The Agricultural Scientific Enterprise: A System in Transition, ed. L. Busch and W. B. Lacy. Boulder, CO: Westview Press.

Magrath, C. P. 1994a. Perspective. NASULGC Newsline, May, pp. 5–6.

———. 1994b. The twenty-first century land-grant university. Unpublished paper presented at the Conference on Food System Professional Education Initiative, sponsored by the W. K. Kellogg Foundation, 8 June, Olive Branch, MS.

Malone, T. F. 1994. Sustainable human development: A paradigm for the twenty-first century, challenge and opportunity for higher education. White paper for the National Association of State Universities and Land Grant Colleges, 6 November.

Manderscheid, L. V. 1988. Undergraduate educational opportunities in the face of declining enrollment. American Journal of Agricultural Economics 70:985–993.

Mathews, D. W. 1990. Allocating resources as a means for inducing and responding to change. In Managing Change in Higher Education. New Directions in Higher Education, no. 71, ed. D. W. Steeples. San Francisco: Jossey-Bass, pp. 37–50.

Maw, I. L. 1994. The need for future perspectives on academic programming in the land-grant institution. Unpublished paper presented University of Connecticut, Storrs, 18 January.

Mawby, R. G. 1985. Paper presented to deans and directors of land grant colleges of agriculture. North Central Region, Chicago, IL.

Merritt, R. H., and K. Wilson. 1990. University review: Key questions using systems techniques. NACTA Journal 34:14–19.

Meyer, J. H. 1992. Rethinking the Outlook of Colleges Whose Roots Were in Agriculture. Davis, CA: University of California Press.

———. 1993. The stalemate in food and agricultural research, teaching, and extension. Science 260:881–882.

———. 1997. Re-engineering the Land Grant College of Agriculture. Davis, CA: University of California Press.

———. 1998. The Historical Trek of the Land Grant College of Agriculture. Davis, CA: University of California Press.

Miller, F. P. 1995. Forces driving changes in colleges of agriculture. Journal of Natural Resources and Life Sciences Education 24:9–13.

Mills-Packo, P. A., K. Wilson, and P. Rotar. 1991. Highlights from the use of the soft systems methodology to improve agrotechnology transfer in Kona, Hawaii. Agricultural Systems 36:409–425.

Naisbitt, J. 1982. Megatrends. New York: Warner Books.

Naisbitt, J., and P. Aburdene. 1990. Megatrends 2000. New York: William Morrow.

National Research Council (NRC). 1988. Educating the Next Generation of Agricultural Scientists. Washington, DC: National Academy Press.

———. 1992. Agriculture and the Undergraduate. Washington, DC: National Academy Press.

———. 1996. Colleges of Agriculture at the Land Grant Universities: Public Science and Public Policy. Washington, DC: National Academy Press.

Newton, C., and T. Tarrant. 1992. Managing Change in Schools: A Practical Handbook. New York: Routledge.

Nipp, T. L. 1988. Congress and the future of agricultural research, extension, and education. Journal of Production Agriculture 1:187–188.

North Central Region Curricular Committee Project. 1987. Curricular innovations for 2005: Planning for the future of our food and agricultural sciences. University of Wisconsin, Madison.

———. 1989. Educating for a global perspective: International agricultural curricula for 2005. University of Wisconsin, Madison.

Offutt, S. E. 1993. The National Academy of Sciences study of agriculture programs in land grant universities. Towards the Twenty-first Century: A Multidimensional Transition of the State Agricultural Experiment Stations. National Futuring Conference, Experiment Station Committee on Organization and Policy, 15–16 June, Washington, DC, pp. 47–49.

Pacey, L. 1993. Collaboration is the key to success: Lessons learned from a comprehensive education program. Unpublished paper presented at the Ninth Annual Conference on Teaching and Learning at a Distance, 4–6 August, Madison, WI.

Pascarella, E. T., and P. Terenzini. 1991. How College Affects Students: Findings and Insight from Twenty Years of Research. San Francisco: Jossey-Bass.

Pennock, R., Jr., and D. C. Scanlon. 1985. Identification of agronomic competencies essential for graduates of baccalaureate degree programs. Journal of Agronomic Education 14:93–97.

Perry, W. G., Jr. 1981. Cognitive and ethical growth: The making of meaning. In Arthur W. Chickering and Associates, The Modern American College: Responding to the New Realities of Diverse Students and a Changing Society. San Francisco: Jossey-Bass.

Poley, J. 1999. Ten questions, big and small, about the impact of technology in contemporary society. Journal of Family and Consumer Sciences 91(3):38–40.

Pulver, G. C. 1994. Restructuring in rural America. Unpublished paper presented at the Third Annual National Symposium for Agricultural Finance Executives, 18 April, St. Louis, MO.

Reich, R. R. 1991. The Work of Nations. New York: Alfred A. Knopf.

Richards, D., and S. Engle. 1991. After the vision: Suggestions to corporate organization. California Management Review.

Rush, S. C. 1992. Productivity or quality? In Productivity in Higher Education, ed. R. E. Anderson and J. W. Meyerson. Princeton, NJ: Peterson's Guides, pp. 1–13.

Ruttan, V. W. 1991. Moral responsibility in agricultural research. In Beyond the Large Farm: Ethics and Research Goals for Agriculture, ed. P. B. Thompson and B. A. Stout. Boulder, CO: Westview Press, pp. 107–124.

Schillo, K. K. 1999. An appropriate role for ethics in teaching contemporary issues. Journal of Animal Science 77(Supplement 2):154–162.

Schuh, G. E. 1986. Revitalizing land grant universities: It's time to regain relevance. Choices 1:6–10.

———. 1993. The future of land grant universities. Journal of Agricultural and Applied Economics 25:112–121.

Siebert, L. A. 1982. The Survivor Personality. Book manuscript. Portland, OR.

Simmons, S. R., R. K. Crookston, and M. J. Stanford. 1992. A case for case study. Journal of Natural Resources and Life Science Education 21:2–3.

Society of American Foresters (SAF). 1992. Forest resource management in the twenty-first century: Will forestry education meet the challenge. Society of American Foresters, Bethesda, MD.

Stauber, K. 1993. Alternative visions for the next generation land-grant university. Towards the Twenty-first Century: A Multidimensional Transition of the State Agricultural Experiment Stations. National Futuring Conference, Experiment Station Committee on Organization and Policy, 15–16 June, Washington, DC, pp. 50–57.

Steeples, D. W., ed. 1990. Managing Changes in Higher Education. New Directions in Higher Education, no. 71. San Francisco: Jossey-Bass.

Thompson, P. B., R. J. Matthews, and E. O. van Ravenswaay. 1994. Ethics, Public Policy, and Agriculture. New York: Macmillan.

Thompson, R. P., O. Capps Jr., and J. C. Massey. 1994. Demand for an undergraduate education in the agricultural sciences. American Journal of Agricultural Economics 76:303–312.

Toffler, A. 1970. Future Shock. New York: Random House.

Townsend, C. D. 1992. Be prepared to adjust: Team leadership is here! Association of Leadership Educators, Summer.

Tucker, A. 1984. Chairing the Academic Department: Leadership Among Peers. 2d ed. New York: Macmillan.

Turner, C. M. 1990. Responding to change: The need for flexible college structures and practices. Mendip Papers, Staff College, Bristol, England, p. 14.

Vietor, D. M., and H. T. Cralle. 1992. Value-laden knowledge and holistic thinking in agricultural research. Agriculture and Human Values 9(3): 44–57.

Vietor, D. M., and R. F. Lucey. 1978. A cropping problem in undergraduate agronomy courses. Journal of Agronomic Education 7: 31–35.

Ward, D. 1994. Technology and the changing boundaries of higher education. Education Review (Jan/Feb): 23–27.

Williams, P. H., and H. L. Young. 1992. Teaching science as inquiry. In National Research Council, Agriculture and the Undergraduate. Washington, DC: National Academy Press, pp. 204–207.

Wilson, K., and G. E. B. Morren Jr. 1990. Systems Approaches for Improvement in Agriculture and Resource Management. New York: Macmillan.

Wingspread Group on Higher Education. 1993. An American Imperative: Higher Expectations for Higher Education. Racine, WI: Johnson Foundation.

Wood, R. J. 1990. Changing the educational program. In Managing Change in Higher Education. New Directions in Higher Education, no. 71, ed. D. W. Steeples. San Francisco: Jossey-Bass, pp. 51–58.

Index

ISBN 0-8138-1774-9
90000
9 780813 817743